MW01708675

Planning Occupational Health & Safety

Important Disclaimer

No person should rely on the contents of this publication without first obtaining advice from a qualified professional person. This publication is sold on the terms and understanding that (1) the authors, consultants and editors are not responsible for the results of any actions taken on the basis of information in this publication, nor for any error in or omission from this publication; and (2) the publisher is not engaged in rendering legal, accounting, professional or other advice or services. The publisher, and the authors, consultants and editors, expressly disclaim all and any liability and responsibility to any person, whether a purchaser or reader of this publication or not, in respect of anything, and of the consequences of anything, done or omitted to be done by any such person in reliance, whether wholly or partially, upon the whole or any part of the contents of this publication. Without limiting the generality of the above, no author, consultant or editor shall have any responsibility for any act or omission of any other author, consultant or editor.

Planning Occupational Health & Safety

CCH
a Wolters Kluwer business

a guide to OHS risk management

7th Edition

CCH AUSTRALIA LIMITED
GPO Box 4072, Sydney, NSW 2001

Head Office North Ryde
Phone: (02) 9857 1300 Fax: (02) 9857 1600

Customer Support
Phone: 1 300 300 224 Fax: 1 300 306 224
www.cch.com.au

Book Code: 3972A

ABOUT CCH AUSTRALIA LIMITED

CCH Australia Limited is part of a leading global organisation publishing in many countries.

CCH began business in Australia in 1969 and quickly established a solid reputation as the country's leading tax publisher. Today our scope is much broader, and our product range is constantly being expanded to meet customer needs.

CCH publications cover a wide variety of topical areas, including tax, company law, accounting, contract law, conveyancing, human resources, industrial law, occupational health and safety, torts and training. CCH is committed to identifying and integrating new technologies into its products.

Our clients include barristers, solicitors, accountants, human resources managers, OHS specialists, business people and students. We aim to provide up-to-date, accurate, authoritative, knowledge-based, practical information which customers can easily and quickly apply to their own specific circumstances.

Enquiries are welcome on 1300 300 224.

To know more about CCH Australia Limited, visit www.cch.com.au.

National Library of Australia Cataloguing-in-Publication Data

Planning occupational health & safety: a guide to OHS risk management.

7th ed.

Includes index.

ISBN 1 921022 99 X.

1. Industrial safety – Australia. 2. Industrial hygiene – Australia. 3. Personnel management – Australia – Planning. I. Grammeno, Gabrielle. II. CCH Australia Limited.

363.110994

© 2006 CCH Australia Limited

First published .... February 1983

2nd edition .... February 1987

3rd edition .... December 1991

4th edition .... November 1996

5th edition .... June 2000

Reprinted with revisions .... April 2001

6th edition .... January 2004

7th edition .... January 2006

This publication was first entitled *Planning Occupational Safety & Health.*

All rights reserved. No part of this work covered by copyright may be reproduced or copied in any form or by any means (graphic, electronic or mechanical, including photocopying, recording, recording taping, or information retrieval systems) without the written permission of the publisher.

Printed in Australia by McPherson's Printing Group.

FOREWORD

Planning Occupational Health & Safety — a Guide to OHS Risk Management — 7th edition is a practical guide for occupational health and safety (OHS) professionals or others with workplace responsibilities for health and safety, students of the subject, and people employed in human resources departments.

Occupational health and safety is now accepted as an integral part of business operation, due to growing recognition that risks to health and safety need to be managed, along with other areas of possible risk, such as quality management, the environment and customer relations. A planned approach to OHS is recognised as essential, considering the high cost of work-related injury and illness to organisations, individuals and society at large.

This publication aims to describe how to introduce and maintain the OHS function in an organisation. The book includes discussion of the impact of legislation, OHS management systems, factors to consider when planning and implementing the function, the role of various people in the organisation, administration and methods of risk management and review.

Chapter 1 examines the importance of planning OHS, including the cost of unsafe practices and the changes in approach occasioned by these costs.

Chapter 2 provides an overview of the legislation dealing with health and safety in Australia.

Chapter 3 is an outline of the risk management process, which is the basis of the modern approach to OHS legislation. It provides practical advice on tackling each stage.

Chapter 4 outlines the use of OHS management systems, and Chapters 5–7 discuss the infrastructure, planning and practicalities of the risk management process. This includes an assessment of the organisation's current position, staffing considerations, the role of OHS representatives and committees, and techniques for identifying, assessing and controlling health and safety risks and investigating accidents at the workplace. Ongoing evaluation of OHS practice is also discussed, and Chapter 8 outlines the key features of injury management and workers compensation.

Chapters 9–11 provide an introduction to examples of common health and safety problems, setting out some typical approaches to risk control in these areas.

Appendices A, B and C outline further resources to assist with health and safety management, including government departments, private organisations, OHS professionals and sources of information.

For more detailed discussions on any of the topics raised in the book, see the extensive range of OHS products published by CCH Australia Limited.

Information on the state of the art in OHS practice (including information on conferences/seminars, videos, books and education) is published in CCH's bi-monthly magazine, *The Journal of Occupational Health and Safety — Australia and New Zealand*.

CCH also publishes an online reporting service (also available in three volumes of hard copy), *Australian Occupational Health & Safety Law*, which provides the text of OHS legislation in each state and territory. CCH's single-volume loose-leaf service, *Managing Occupational Health & Safety* (also available online), is a practical on-site reference for those managing health and safety. For a more extensive list of CCH titles in the area of OHS, refer to Appendix B.

CCH Australia Limited

October 2005

ACKNOWLEDGEMENTS

CCH Australia Limited acknowledges the contribution to the previous editions of this book by Richard M Ridout, FSIA, Leigh Deves, BA (Hons), MBA, Glyn Williams and Joy Window, BSc (Hons), GradDipEditPub.

This edition was updated by Gaby Grammeno, BA (Hons), MPH.

Editor

Gaby Grammeno

Sub editor

Syama Ramasamy

Book coordinator

Dora Beech

Production editor

Seema Saseedharan

Indexer

Mark Southwell

CONTENTS

Chapter 1

BEST PRACTICE OCCUPATIONAL HEALTH AND SAFETY MANAGEMENT

¶1-100 Introduction

Every year, around two million people lose their lives because of work-related injury and illness. This means an average of over three people dying every minute, due to on-the-job health and safety problems. According to the World Health Organization's global estimates, about 1.7 million of these deaths are due to occupational diseases, for example, cancer or other illness, caused by exposure to workplace hazards such as chemical substances or asbestos. It has often been noted that the occupational death toll is higher than that for road accidents or war.

Apart from cancer, the most common job-related health problems are respiratory diseases, musculoskeletal disorders, hearing loss, circulatory diseases and communicable diseases caused by exposure to pathogens. Worldwide, asbestos alone is reported to be responsible for 100,000 deaths per year. In addition to the deaths, literally hundreds of millions of people are hurt in non-fatal work accidents each year.

Health and safety failures at work cost Australia at least $31 billion per year in lost production, treatment of injuries and illnesses, and rehabilitation and compensation for those injured or made sick by their work, according to estimates from the Productivity Commission's 2004 report, *National Workers' Compensation and Occupational Health and Safety Frameworks*. This imposes very significant costs on individuals, businesses, the community and the economy as a whole.

These facts highlight the challenges faced by employers in Australia, in fulfilling their legal obligations to provide safe and healthy working conditions.

¶1-120 Changing approaches to occupational health and safety

In past decades, occupational health and safety (OHS) law in Australia consisted of thousands of pages of detailed requirements specifying restrictions on work practices. For example, legislation applying to factories, shops and industries prescribed that all dangerous machinery must be guarded, and women could not legally be asked to lift weights of more than 16 kg.

Employers were expected to comply with prescriptive requirements such as these, and it was hoped that these measures would protect the health and safety of people at work.

By the mid 1970s, however, it was realised that this approach, known as "minimum standards compliance", was very limited in its ability to protect people from work-related injury and illness. Much of the workforce was not covered by the legislation. With regard to back injuries, for example, the restrictions on weights to be lifted by women and juniors in factories gave no protection to women working as nurses in hospitals or as carers in childcare centres, and no protection at all to men. Moreover, many injuries to backs, shoulders and necks occurred as a result of falls, repetitive movements or awkward postures workers were obliged to adopt because of poorly designed equipment or workplaces — in other words, the problem could not be fixed merely by restricting the weight of loads to be lifted.

This realisation led to a new approach to controlling health and safety problems. In a series of changes to health and safety laws in all Australian states and territories, legislation focusing on outcomes — that is, on employers' performance in protecting workers' health and safety — was introduced across Australia in the 1980s and 1990s. The new laws required employers to ensure their workers' health and safety was not endangered by their work, as far as is reasonably practicable.

This shift to so-called "performance-based" legislation has changed the way employers must deal with threats to health and safety. It is no longer considered adequate or acceptable merely to ensure that workers have suitable workers compensation cover, provide first aid facilities and exhort workers to be careful. Employers' duty of care to workers and others affected by their undertakings has been written into Acts of parliament in all Australian jurisdictions, and the process of meeting legal requirements and efforts in harm minimisation have become more complex and sophisticated. Risk management has become the key tool of OHS practitioners in establishing prevention priorities, given the vast array of potential risks to be mitigated to make the workplace "healthy and safe" and to demonstrate due diligence.

There has also been a shift in what organisations expect from their health and safety programs. Cost — in the context of organisational goals relating to competitiveness — has become more of a concern, although even today few organisations measure and monitor the direct and indirect costs of workplace incidents in the same way that injuries and incidents are monitored.

Benefits of best practice health and safety management

While there is plenty to be done for many organisations within a minimum compliance/harm minimisation framework, best practice organisations have moved to embrace a broader set of goals and objectives.

Within this framework, companies recognise that organisational health and safety in its broadest sense can positively contribute to a company's strategic objectives. Companies can benefit from positively promoting the health and wellbeing of employees — organisational design that meets the needs of those that work within it will support more efficient, effective production.

Moreover, organisational goals for best practice organisations extend beyond short-term profitability and cost minimisation. Corporate social responsibility recognises that the value of reputation in the market place and the growth in socially responsible investment funds (and associated screening for health and safety performance) makes even traditional OHS a potential business development opportunity (or risk) rather than a minor internal function.

As well as these considerations, developments such as the increase in terrorist attacks and the outbreaks of SARS (Severe Acute Respiratory Syndrome) and "bird flu" have led many organisations to question their readiness for managing emergency situations and to ask whether the impact on their business continuity would be economically catastrophic. In addition, high profile financial failures, such as insurers HIH in Australia and energy company Enron in the United States, highlighted the importance of corporate governance and the system of internal controls aimed at mitigating financial risks.

The impact of these types of events on investor confidence (as evidenced by stock market falls worldwide) has been substantial. Investors want to be assured that companies are effectively managing risks.

> This book promotes an integrated approach to occupational health and safety. It puts OHS in the context of a wider range of risks which might impede the organisation's capacity to meet its objectives. Health and safety is seen as the concern of the whole organisation, with a particular emphasis on management's role in planning and supporting the management of OHS risks. Risk management becomes the focus, rather than compliance with regulations or reaction to individual occurrences.

The purpose of this chapter is to indicate the magnitude of the health and safety problem for organisations and to show why a planned approach to health and safety is essential. This introduction includes explanations on the role of management and employees in a planned health and safety program, and the aims of such a program.

¶1-130 Measuring the problem

In Australia the problem is usually measured by considering the number of workers compensation cases. According to the National Occupational Health and Safety Commission's *Compendium of Workers' Compensation Statistics Australia*, in 2001–02 there were 38,810 accepted workers compensation cases which resulted in a fatality, or a permanent or temporary disability involving a week or more off work.

The actual number of work-related injuries and illnesses would be considerably higher, according to the *National Workers' Compensation and Occupational Health and Safety Frameworks*. In support of this claim, the productivity commission cited a recent survey by the Australian Bureau of Statistics which found that many individuals who experienced a work-related injury or illness did not apply for workers compensation. In most cases this was because the injury was considered to be minor, but other reasons included a lack of awareness of eligibility or the availability of benefits, the negative impact on employment, the effort of making a claim, or the employer agreeing to pay the cost outside a workers compensation scheme.

Studies of coroners' reports have confirmed that the 300 or so occupational fatalities reflected annually in workers compensation statistics do not tell the whole story. The real number of work-related deaths is much higher, as many are never the subject of workers compensation claims.

In addition, many occupational diseases are never recognised as being wholly or partly attributable to work, therefore these too, are not reflected in workers compensation statistics. The National Occupational Health and Safety Commission estimated that there are about 2,200 deaths a year due to workplace exposure to hazardous substances such as asbestos and harmful chemicals.

Diseases and injuries experienced by workers such as sole traders and independent contractors are also omitted from official statistics, as most of these are not covered by workers compensation. These limitations on the data mean that the problem is in fact much larger than it may appear from workers compensation figures.

The scope of the impact

While it is difficult to assess accurately the size of the problem, it is clear that the cost to the Australian community is huge. As well as the direct costs to individuals in terms of distress and disruption, business bears a burden of hidden, indirect costs — discussed at ¶1-140. Other hidden costs such as rehabilitation centres and other hospitalisation costs also add to the total burden on the community.

In the light of Australian industry's strivings for international competitiveness, there is substantial potential for cost reductions in this area. Costs can be reduced by eliminating risks as far as practicable, by reducing the severity and the number of work-related injury and disease cases, and by better management of injury cases which do occur.

The human factor

Besides the "drama" of death or personal injury, there is a hidden incidence of occupational disease, which in many cases is a slow deterioration going unnoticed in the early stages. Examples include occupational hearing loss, dust diseases and occupational cancer. It may be only after a time delay of five years or more that the existence of such diseases is established, by which time their effects may have become irreversible.

The social costs of work-related injury (to the affected employee, his/her family and the community) can be less obvious than the economic impact. An employee whose injury disallows continuing employment becomes to some extent invisible and, once the individual matter of compensation is settled, the incident may no longer be an influence for change at that or other workplaces.

¶1-140 Indirect costs

There can be no dispute that injuries, ill-health and deaths of workers represent very significant insurance costs to an organisation, but on top of these direct insurance costs and the basic human considerations are other losses of which management should be particularly aware. These include indirect costs such as loss of productivity, loss of morale, labour turnover, absenteeism, industrial disputation over working conditions, and an impaired public image which may be reflected in loss of orders/accounts.

The costs to management are often divided into "insured" and "uninsured" categories. Note that, although the "uninsured" costs are in many instances difficult to calculate, their potential is to create a total cost several times that of the total insured costs. The following table shows some types of costs associated with accidents.

Costs of accidents

INSURED COSTS	UNINSURED COSTS
Injuries • Compensation for lost earnings • Medical and hospital costs • Awards for permanent disability • Rehabilitation costs • Funeral charges • Pensions for dependants • Public liability ***Property damage*** • Fire • Loss and damage	***Injuries*** • First aid expenses • Transportation costs • Cost of investigations • Cost of processing reports ***Wage losses*** • Idle time of workers whose work is interrupted • Person hours spent in cleaning up the accident area • Time spent repairing damaged equipment • Time lost by workers receiving first aid • Possible extra overtime ***Production losses*** • Product spoiled by accident • Loss of skill and experience • Lowered production of worker replacement • Idle machine time • Unfavourable public image • Loss of goodwill ***Associated costs*** • Difference between losses and amount recovered • Rental of equipment to replace damaged equipment • Recruitment of workers for replacement of injured employee • Training of replacement worker • Wages or other benefits paid to disabled worker • Overhead costs while production is stopped • Loss of bonus or payment of forfeiture for delays

INSURED COSTS	UNINSURED COSTS
	Off-the-job accidents
	• Cost of medical services
	• Time spent on injured worker's welfare
	• Decreased production of replacement
	Intangibles
	• Raised staff turnover
	• Lowered employee morale
	• Increased labour conflict

An organisation with an active, ongoing, consultation-based health and safety program will have reduced the risk of accidents occurring and have efficient procedures to deal with the human or technical emergencies that become part of a work accident. Money, time and, most importantly, lives can be saved. The data gathered through accident/incident investigation can be fed back into the overall management administration to help prevent future incidents.

Accidents may also have harmful effects on the general environment beyond the workplace. For example, if an oil barge discharges oil into a river, there may be damage to the foreshores, other boats, wildlife and nearby homes. Pollution from chimney discharge may affect nearby residents, cars parked in the street, etc. Management should also be aware of the implications of public liability (persons other than employees who become injured) and product liability (adverse consumer reaction, loss of sales, cost of recalls, etc).

¶1-150 Performance-based requirements

The "cost" realisations referred to at ¶1-140, as well as the recognition that the old style of legislation was inadequate, have prompted changes in the approach to the practice of occupational health and safety in Australia. These changes emphasise practical means to achieve injury prevention and to encourage health maintenance and enhancement at the workplace. The changes include the following:

1. The traditional "minimum standards"/compliance approach to the formulation of the law on health and safety is an inheritance from an earlier, more static, work environment and is ill-suited to keeping pace with the rapid and uneven adoption of new work practices initiated by technological change. This has given way to a "performance-based"/ self-regulated approach which emphasises the employer's duty of care and a process of risk management (detailed in Chapter 3), as well as consultation with employees (outlined in Chapter 2). The term "performance-based" as it is applied to the modern style of health and safety legislation draws on the notion that any system of protecting workers from work-related injury or illness is only as good as its performance — that is, the outcome is of primary importance, rather than compliance with minimum standards.
2. Increasingly, codes of practice provide detailed instructions on how certain industries can operate safely, how to avoid certain common

injuries (such as Occupational Overuse Syndrome, manual handling injuries and occupational hearing loss) or how to handle certain dangerous substances (such as asbestos, vinyl chloride or synthetic mineral fibres). National codes of practice produced by NOHSC are merely advisory, while some states have chosen to adopt codes of practice in their legislation.

3. All states and territories have revamped their workers compensation schemes, which continue to be volatile, in their attempts to deal with spiralling workers compensation costs.

4. Historically, occupational health and safety has been regulated by separate legislation in each state and territory, with the result that in Australia there are nine separate sets of health and safety laws (for the six states, two territories and the Commonwealth). The establishment of the National Occupational Health and Safety Commission (NOHSC) in 1985 created a gradual movement towards more uniform requirements. This is reflected now in a strong degree of consistency Australia-wide in many key requirements, though differing details mean that reference should always be made to local state/territory legislation. The NOHSC has been replaced by the Australian Safety and Compensation Council (ASCC). Like its predecessor, the new body is comprised of representatives from the Commonwealth government, as well as each state and territory government, along with employer and employee representatives, and it will continue NOHSC's role in developing national OHS standards and codes of practice. The ASCC's main aim is to establish a national approach to workplace safety and workers compensation, and it will also coordinate research and provide policy advice to the Workplace Relations Ministerial Council — which comprises the Federal Workplace Relations Minister and state and territory counterparts.

¶1-160 Management's role

The main emphasis of this book is that the dangers of the workplace cannot be dealt with haphazardly; risks must be eliminated, or at least controlled, through the basic functions of risk management, in the context of planning, organising, staffing and controlling.

Management's role in this process should be clearly seen and demonstrated. The employment of trained OHS personnel, the establishment of health and safety representatives and/or committees and cooperation with these, as well as the provision of adequate financial backing for their function, will, of course, be management's responsibility. Most importantly, management at all organisational levels needs to demonstrate by personal example and positive attitudes, in addition to its financial support, a commitment to high standards of health and safety for all employees wherever they may be employed. The standard of care demanded of an employer is high and one owed to individual workers rather than to workers in general.

If workers are not made aware of the reasons for the implementation of an OHS risk management program or do not see management staff complying with it, they will become suspicious of the motives behind such a program. They may resent any interference with their established work patterns, and may not report certain accidents or incidents if they feel they are being monitored, or that they could be blamed or punished for reporting them. Therefore, they must be convinced that change is essential before they will accept it.

Workers should be involved in the planning and implementation of health and safety programs (see ¶1-170). Yet it is management which must accept final responsibility in regard to its workers, its shareholders, its legal requirements, and the community at large. It is only on the basis of management responsibility and accountability that OHS programs will succeed.

Strategic planning versus operational planning

Management will be concerned with ensuring that the organisation's overall planning (strategic planning) incorporates health and safety. This will involve giving health and safety a high priority, staffing and financing the health and safety function, and determining what safety issues should receive priority.

In relation to health and safety, operational planning (which is concerned with shorter-term matters) will be carried out by supervisors or the human resources department (see ¶1-180) in consultation with employees and with the involvement of OHS representatives or committees, as appropriate.

¶1-170 Employees' role — participation and the consultative approach

The modern approach to occupational health and safety emphasises consultation between employers and employees. The usual method of achieving this is through health and safety representatives and/or a health and safety committee. In all states and territories, there are legislative requirements to elect/appoint either representatives or a committee (see ¶2-140).

Health and safety representatives are generally elected by the employees, with the workplace being broken down into locations/units, and each location/ unit electing one representative. Elected representatives will generally be more acceptable to employees than appointed representatives. Health and safety committees may be made up of elected representatives and appointed employer representatives. In each state, the legislation provides some framework as to how the representatives and/or committee members must be selected (see ¶2-140).

The functions of representatives and committee members are discussed at ¶5-470.

In addition to consultation through representatives and committees, employees should be made aware of all health and safety initiatives at the workplace through meetings, circulars and supervisor communication. Employees should also be made aware of the results of all health and safety programs.

Advantages of a consultative approach

Involving workers in health and safety will have benefits in terms of both the information that will be available to the health and safety program and the commitment of the workers to the program.

With respect to the additional information workers can provide, they are the closest to the work process and therefore more aware of what methods, procedures and practices apply, which machines have certain problems or peculiarities, the types of minor incidents which occur frequently and the short cuts which are practised to complete jobs on time. Information such as this is invaluable to a health and safety program. Workers are also best placed to monitor the effectiveness of a program after it is introduced.

Equally invaluable to the success of any program is the cooperation and commitment of the workers. These can be enhanced by consultation on workplace safety issues. Genuine consultation will demonstrate to workers that the program is a real attempt to improve their safety, rather than an exercise in minimum standards compliance. Workers who become involved in the consultation process (that is, those who become representatives or committee members) can be expected to greatly increase the awareness of their own safety, and the safety of their immediate section.

¶1-180 The role of the human resources/health and safety department

It has already been noted that the commitment to health and safety must come from the top of an organisation. However, the day-to-day resourcing of the health and safety function may be carried out with the assistance of a human resources department. In large organisations, a specific health and safety department may be constituted.

Such day-to-day resourcing may involve a wide range of functions, including assistance with the following:

- identifying hazards to health and safety, and assessing the risks arising from those hazards
- controlling the risks to health and safety, and monitoring the success of the control measures
- implementing specific programs
- preparing feedback to management on the effectiveness of programs
- liaising with health and safety professionals and/or medical professionals
- administering health and safety committee meetings or meetings between representatives and management

- recording all unsafe incidents and recommending changes to prevent recurrence of such incidents
- performing regular inspections or arranging for professionals and worker representatives to perform inspections
- maintaining first aid equipment and a first aid room
- monitoring the condition of any safety equipment, and
- compiling and keeping up to date all safety information, including material safety data sheets.

Those involved in managing health and safety may encounter a certain resistance to new safety practices, perhaps as part of a general resistance to change. Disputes can best be avoided through proper consultation and education.

¶1-190 Aims of an OHS risk management program

The aim of an OHS risk management program is to identify and remove or control the causes of accidents and health hazards at the workplace. This will involve an assessment of the current safety and health position at the workplace, correction of any hazards, and the development and documentation of safe work methods to a satisfactory standard with the establishment of health and safety education, promotion and training programs.

The goals of such a program will be best achieved through the involvement of both workers and management in some form of health and safety organisation (such as health and safety committees or health and safety representatives) and through the employment of health and safety personnel (such as OHS managers). Such staffing matters are discussed in Chapter 5. Setting up a health and safety program is discussed in Chapter 6.

¶1-200 Some definitions

Before dealing with the process of developing an OHS program, it would be helpful to clarify a few of the terms used in OHS.

Occupational health

First of all, the basic concept of "occupational health" needs to be considered. Several definitions of occupational health and safety, and occupational health services have been produced by professional bodies, international organisations such as WHO and ILO, and national bodies and authorities. If one summarises those definitions, occupational health is considered to be multi-disciplinary activity aiming at:

- protection and promotion of the health of workers by preventing and controlling occupational diseases and accidents, and by eliminating occupational factors and conditions hazardous to health and safety at work
- development and promotion of healthy and safe work, work environments and work organisations

- enhancement of physical, mental and social wellbeing of workers and support for the development and maintenance of their working capacity, as well as professional and social developments at work, and
- enablement of workers to conduct socially and economically productive lives and to contribute positively to sustainable development.

The concept of "occupational health" is, then, a positive one, which focuses on the need for maintenance of good health and wellbeing. Of course, OHS programs will deal with the treatment of ill-health as well, but prevention should be the primary motive in any program.

Occupational health and safety matters include:

(a) the physiological and psychological needs and wellbeing of persons engaged in occupations

(b) work-related death

(c) work-related trauma

(d) the prevention of work-related death or work-related trauma

(e) the protection of persons from, or from risk of, work-related death or work-related trauma

(f) the rehabilitation and retraining of persons who have suffered work-related trauma.

Hazard

A hazard is a source or a situation with a potential for harm in terms of human injury or ill-health, damage to property, damage to the environment, or a combination of these.

Risk

A risk is often thought of as the danger arising from the presence of a hazard. It is the combination of the likelihood, or probability of occurrence, and the consequences, of a specified hazardous event.

Hazard identification

Hazard identification is the process of recognising a hazard exists and defining its characteristics. This process is referred to as "risk identification" in the relevant Australian Standard, AS 4360, but as "hazard identification" in much Australian OHS legislation.

Risk assessment

Risk assessment is the overall process of estimating the magnitude of risk and deciding what actions will be taken.

Risk control

Risk control is the process of eliminating or minimising risks.

Risk management

Risk management is the whole process comprising hazard identification, risk assessment and risk control, including monitoring risk controls for effectiveness and adjusting them if necessary.

Occupational rehabilitation

Occupational rehabilitation is the managed process of maintaining injured or ill employees in, or returning them to, suitable employment.

Injury management

This term encompasses all the activities associated with ensuring the safe and early return to work of an injured worker to the workplace. "Workplace injury management" is a broader term than "occupational rehabilitation". It includes treatment, claims management and employment practices, as well as rehabilitation.

Ergonomics

Planning a safe, healthy and comfortable working environment necessitates a basic understanding of the concept of ergonomics. Ergonomics may be defined as the scientific study of the physical relationship between people, the equipment they use and their general environment. As an applied science, ergonomics is involved with the design of equipment and working environments that enable the best use of human capabilities without exceeding human limitations. (In the US, ergonomics is usually referred to as human factors engineering.)

To some extent the term "ergonomics" has become debased through an insufficient understanding of the problems to which it is applied and the mislabelling of half-considered "solutions" as "ergonomic". A striking example of this can be seen in the often costly purchase of "ergonomically designed" furniture as a solution to problems encountered with the introduction of computer technology. While the design of some furniture can contribute to postural problems, more appropriately designed furniture does not prevent a person adopting an inefficient and potentially damaging working posture, when it is not used properly.

Ergonomics cannot rightly be considered as a single subject but rather as a *principle* that should be involved in all aspects of work design and planning. For instance, ergonomic considerations are involved in subjects as diverse as the design of hand tools, the positioning of machine controls and the layout of workplaces. It also includes highly specialised studies such as the application of psychology to the ability of the brain to absorb and use information from display screens or the capacity of the body to work under extreme conditions of temperature.

Accidents and incidents

A workplace accident may be described as any event arising out of employment which results in a work injury or damage to property, or creates the possibility of injury or damage.

Near misses (incidents) and property damage should be taken into account in any accident prevention program as these events give warning that there is something in the work process that needs to be investigated and possibly changed before a serious accident occurs.

¶1-210 Scope of this book

Planning Occupational Health & Safety — a Guide to OHS Risk Management is written against a background of increasing emphasis on risk management in occupational health and safety.

Where relevant, reference is made to the impact of specific legislation, but generally the approach taken in this book is not legalistic but that of a guide to planning. Subsequent chapters of this book outline and discuss the issues and procedures involved in administering the safety and health function, as well as the workers compensation and rehabilitation functions of an individual workplace.

Topics covered include health and safety legislation, risk management, OHS management systems, staffing issues, policy, planning, implementation and evaluation of OHS risk management programs, incident investigation, recording and reporting, and managing workers compensation and occupational rehabilitation. Also provided is an overview of management strategies with regard to particular health and safety problems.

Chapter 2

OVERVIEW OF LEGISLATION

¶2-100 Purpose of this chapter

As mentioned in Chapter 1, Australian occupational health and safety laws have adopted a performance-based, consultative approach grounded in the principles of duty of care and risk management. The purpose of this chapter is, therefore, to set out the legislative framework within which occupational health and safety operates.

Summary of types of legislation

There are various forms of legislative provisions relevant to occupational health and safety. The following summary is expanded upon in the remaining paragraphs of this chapter. Note that liability imposed by the common law exists in addition to the legislative provisions such as Acts and regulations (see ¶2-170).

1. *Acts*: The states and territories are responsible for the law and its administration in the area of occupational health and safety. Each state and territory has its own (variously titled) health and safety Act, which applies to practically all workplaces. These Acts require employers to ensure the health and safety of their employees, and also impose broad duties on employees, contractors, the self-employed and others at workplaces. The Acts also create administrative bodies and empower the government to make regulations covering specific areas. Another feature of these Acts is the requirement to establish consultative mechanisms such as health and safety representatives/committees.

   Each state also has its own workers compensation legislation, which provides for those injured in the course of their employment. Requirements to rehabilitate workers are generally contained within workers compensation legislation.

   There are also Acts (statutes) in each state setting out safety requirements for specific industries or areas (for example, coal mining and dangerous goods).

2. *Regulations*: Regulations are legally binding documents setting out duties for various parties with regard to health and safety in the workplace. Once adopted in a state or territory, regulations must be complied with. Employers may be prosecuted for breaching their obligations under them — mandatory requirements in the regulations are indicated by the use of the words "must" and "shall".

   Regulations made under the OHS and associated Acts provide more detailed requirements for specific areas of workplace safety and health. Some regulations will affect all workplaces while others apply to particular industries.

3. *Codes and standards*: Codes of practice are not legislation, but have the backing of government and act as guides to procedure in particular areas. Codes can be used in determining whether an employer has breached its duty to ensure the safety of employees.

   Standards are similarly not legislation (although there are many instances where an Australian standard has been incorporated into regulations). They are published by Standards Australia (see ¶2-160) after consultation with concerned groups, and are accepted as authoritative guides to good practice. The National Occupational Health and Safety Commission has also developed national standards relating to occupational health and safety.

4. *Industrial awards and agreements* are made pursuant to federal or state industrial legislation and are enforceable at law. They often contain provisions relevant to health and safety, such as protective clothing, first aid facilities and regular rest breaks.

   Industrial awards and enterprise bargaining agreements are not considered in any detail in this book. Note that in theory industrial awards and enterprise bargains could deal specifically with health and safety and, where this occurred, they would have to be taken into account, as well as state OHS legislation, in relation to that industry or workplace.

¶2-120 Modern legislative approach

Until the 1970s OHS legislation in Australia was based solely on the specification of standards. The legislation was directly descended from the earliest English Factories Acts. A realisation that this type of legislation was not very effective in protecting workers' health and safety has led to the introduction of a new style of legislation.

Impetus for the new style of legislation came from the report of a UK inquiry into health and safety — the *Robens Report* — released in 1972. The new approach is based on the expression of general duties of care and provisions for the participation of employees, coupled with better administration and inspection of health and safety.

Between 1972 and 1989, all Australian states and territories introduced the newer style of legislation. This is summarised at ¶2-130. However, some of the older, specifications-oriented Acts remain in force. The more important of these are also mentioned at ¶2-130. Legislative review and amendment is a continuous process, however, and changes are expected in relation to some of this legislation in the medium term.

National uniformity and national strategy

National uniformity in OHS legislation is a daunting goal, as a large amount of state and territory legislation presents points of difference which are gradually being worked through and aligned. The goal has yet to be achieved, though there is considerably more uniformity now than there was 15 years ago. The states and territories have progressively adopted legislation based on national standards and model regulations, for example in the areas of manual handling, hazardous substances, plant and certification for operators of industrial equipment.

In May 2002, the Workplace Relations Ministers' Council, comprising all Commonwealth, state and territory ministers with responsibility for OHS policy, unanimously endorsed a 10-year National OHS Strategy. The vision is for Australian workplaces to be free from death, injury and disease and, for the first time, national targets in relation to reductions in work-related injury and death have been set as milestones on that journey.

The national strategy also commits governments, employer and employee organisations to work together to develop, implement and promote increasingly evidence-based OHS interventions in the five national priority areas:

1. reducing high incidence and severity risks
2. developing the capacity of business operators and workers to manage OHS effectively
3. preventing occupational disease more effectively
4. eliminating hazards at the design stage, and
5. strengthening the capacity of government to influence OHS outcomes.

Risk management

Experience has shown that a legislative approach relying on the prescription of minimum standards and enforcement by inspectors is not always effective. It neglects many of the risk factors for work-related injury and illness, and does not provide employers with a clear process for achieving good safety and health standards. In order to prevent work-related illness and injury, modern health and safety legislation requires a risk management process to be carried out.

What is risk management?

Risk management involves a detailed and systematic examination of any activity, location or operational system to identify risks, understand the likelihood and potential consequences of the risks, and to review the possible approaches to controlling the risks. Suitable risk control strategies should be implemented and then checked to ensure they are appropriate and effective.

Successful risk control can include outcomes such as improved safety, health, production, quality, environmental protection and community acceptance. In the context of occupational health and safety, the first stage of the risk management process is often known as "hazard" identification, rather than "risk" identification. Both terms imply a systematic process for spotting threats to health or safety.

Risk management and legislation

The modern approach expressed in legislation and codes of practice for occupational health and safety calls for risk management to underpin any implementation of health and safety measures. See Chapter 3 for details on the risk management process.

¶2-130 Summary of occupational health and safety legislation

The following provides an overview of the principal health and safety legislation in each jurisdiction. Other relevant Acts are simply mentioned.

Note that the principal Acts do not provide exact specifications for employers to follow. Their purpose is to provide general standards for employers to meet, and to allow prosecutions and/or recovery of damages when those standards are not met. Any specifications which exist are set out in regulations or codes of practice made pursuant to the legislation, or in other, more specific legislation.

The principal Acts differ widely in their terms of coverage and provisions. It would be advisable for employers and employee representatives to ensure they have access to up-to-date versions of the Acts applicable to their workplace. The full text of relevant legislation is available from the websites of OHS authorities, or from the Australian Legal Information Institute (www.austlii.edu.au). Printed copies are also available. For example, the loose-leaf reporting service *Australian Occupational Health & Safety Law* (CCH Australia Limited) contains the full text of all principal legislation, with extracts from other legislation.

The following description of legislation is correct as at 4 October 2005. Requirements in the legislation to establish health and safety representatives or committees are discussed separately at ¶2-140. Enforcement of the legislation is discussed at ¶2-150.

OHS ACTS AND AUTHORITIES			
Jurisdiction	**Act**	**Administered by**	**Maximum penalty**
Federal	*Occupational Health and Safety (Commonwealth Employment) Act 1991*	Comcare Australia	$100,000
NSW	*Occupational Health and Safety Act 2000*	WorkCover NSW	$825,000
Vic	*Occupational Health and Safety Act 2004*	WorkSafe Victoria	$943,920
Qld	*Workplace Health and Safety Act 1995*	Workplace Health and Safety (Department of Industrial Relations)	$750,000
SA	*Occupational Health, Safety and Welfare Act 1986*	South Australian WorkCover Corporation	$200,000
WA	*Occupational Health and Safety Act 1984*	WorkSafe Western Australia	$625,000

OHS ACTS AND AUTHORITIES			
Jurisdiction	**Act**	**Administered by**	**Maximum penalty**
Tas	*Workplace Health and Safety Act 1995*	Workplace Standards Tasmania	$150,000
ACT	*Occupational Health and Safety Act 1989*	ACT WorkCover	$250,000
NT	*Work Health Act 1986*	NT WorkSafe	$125,000

Note that in addition to fines, prison sentences and other types of penalties may also apply under some circumstances. For example, in some jurisdictions sanctions may include enforceable undertakings or court-ordered injunctions, and the courts may order offenders to publish statements relating to their offence.

Industrial manslaughter laws are also under consideration in some states and have been introduced in the Australian Capital Territory. The *Crimes (Industrial Manslaughter) Amendment Act 2003* came into effect on 1 March 2004. This Act makes it possible to convict corporate employers for workplace death or serious injury. The death or serious injury would have to be in the course of the worker's employment. There must also be evidence of criminally reckless behaviour causing serious injury or negligent conduct leading to the worker's death.

Senior officers, including directors and other senior decision-makers in the company, can also be prosecuted under the industrial manslaughter provisions. This is not vicarious liability, as the senior officers' reckless or negligent conduct must have caused the death or injury. The maximum fine for companies is $1.25 million and/or other orders the court may impose, including publicising the offence, consequences and penalties, or undertaking community projects up to $5 million. The Act does not impose any additional OHS obligations on employers. As long as employers ensure that they provide safe and healthy workplaces, they will not be subject to prosecution.

Australian Safety and Compensation Council, formerly the National Occupational Health and Safety Commission

The National Occupational Health and Safety Commission (NOHSC) was established to work cooperatively with state and territory governments, the Australian Chamber of Commerce and Industry, and the Australian Council of Trade Unions, on national priorities for improving OHS. It developed standards and promoted legislative uniformity and cooperation in OHS matters across the country. As a statutory body responsible to the Federal Minister for Employment, Workplace Relations and Small Business, it had no direct authority over OHS in the states, territories or the Commonwealth.

In February 2005 the NOHSC was replaced by the Australian Safety and Compensation Council, which will continue NOHSC's role in relation to national uniformity, the national strategy, and the development of national standards and codes of practice. It will also seek to establish a national approach to workers compensation.

Federal

The Commonwealth has no constitutional power to legislate generally in the area of health and safety. The *Occupational Health and Safety (Commonwealth Employment) Act 1991* regulates the health and safety of federal public servants. It is administered by Comcare Australia.

The Act places general duties of care on employers (the Commonwealth or a Commonwealth authority), employees, and manufacturers and suppliers in much the same way as the state Acts discussed below. The Act also provides for the selection of health and safety representatives in Commonwealth workplaces and, in certain circumstances where the number of employees at the workplace is at least 50, for the establishment of health and safety committees.

Other federal legislation relevant to occupational health and safety includes:

- *Industrial Chemicals (Notification and Assessment) Act 1989*: established a scheme whereby all new chemicals imported into or manufactured in Australia are assessed (known as the National Industrial Chemical Notification and Assessment Scheme (NICNAS))
- *Occupational Health and Safety (Maritime Industry) Act 1993.*

New South Wales

The principal Act in New South Wales is the *Occupational Health and Safety Act 2000*, administered by the WorkCover Authority of New South Wales. The Act places a general duty on employers to ensure the health, safety and welfare at work of all their employees (s 8(1)). This is extended to persons other than employees at the workplace (s 8(2)). There is also a duty of care placed on manufacturers and suppliers of plant and substances (s 11), and on employees to take reasonable care for the health and safety of other persons at the workplace (s 20).

Health and safety inspectors are given the power to inspect any workplace (s 50). Regulations made under the Act allow an inspector to issue improvement notices or prohibition notices requiring an employer to remedy a particular safety risk, to issue on-the-spot fines or to initiate prosecutions.

The Act requires employers to consult employees on workplace health and safety, either through the establishment of workplace health and safety committees or representatives, or in other ways as agreed upon by the employer and employees (see ¶2-140).

WorkCover is given the power to formulate industry codes of practice (s 41). Such codes are not in themselves legislation but they may be used in proceedings under the Act to establish that a person failed to comply with his/her duty of care (see ¶2-160).

The NSW Occupational Health and Safety Regulation 2001 maintains many previously existing provisions relating to manual handling, hazardous substances, plant safety, first aid, and so on, and requires employers to undertake risk management.

Other relevant legislation includes:

- *Coal Mines Regulation Act 1982*
- *Coal Mine Health and Safety Act 2002*
- *Dangerous Goods Act 1975*
- *Electricity Safety Act 1945*
- *Smoke-free Environment Act 2000*, and
- *Occupational Health and Safety Amendment (Workplace Deaths) Act 2005.*

Victoria

The *Occupational Health and Safety Act 2004* is the result of legislative reforms designed to provide employers and employees with more support on health and safety, and to address the social and economic cost of work-related deaths, injuries and illness. Most sections of the new Act came into effect in July 2005. It is the principal occupational health and safety legislation in Victoria, and is administered by WorkSafe Victoria, the health and safety arm of the Victorian WorkCover Authority.

The Act imposes general OHS duties on employers, the self-employed, employees, designers, manufacturers, suppliers and others. These general OHS duties require a person, in most instances, to ensure health and safety so far as is reasonably practicable. This requires the person:

- to eliminate risks to health and safety so far as is reasonably practicable, and
- if it is not reasonably practicable to eliminate risks to health and safety, to reduce those risks so far as is reasonably practicable.

This, in turn, requires people to take into consideration:

- the likelihood of a hazard or risk occurring (ie the probability of a person being exposed to harm)
- the degree of harm that would result if the hazard or risk occurred (ie the potential seriousness of injury or harm)
- what the person concerned knows, or ought reasonably to know, about the hazard or risk and any ways of eliminating or reducing that hazard or risk
- the availability and suitability of ways to eliminate or reduce the hazard or risk; and
- the cost of eliminating or reducing the hazard or risk.

The general duties of care imposed by the Act include: the duty of employers to provide and maintain a safe working environment for employees and independent contractors (s 21); the duty of employers and self-employed persons to ensure that others are not exposed to health and safety risks arising from their undertaking (s 23 and 24); the duty of persons who manage or control workplaces to ensure that the workplace, and access to the workplace, is safe (s 26); the duty of manufacturers and suppliers of plant and substances to ensure that the plant or substances are safe for use (s 29 and 30); and the duty of employees to take care of their own safety and the safety of others at the workplace (s 25).

The Act also provides a mechanism for resolving health and safety issues (s 73). The employer and the health and safety representative should attempt to resolve the issue, with either having the power to direct that work shall cease where the threat is immediate, or to require an inspector appointed under the Act to attend the workplace. Where work ceases because of a safety threat, the employer is entitled to assign the employees to suitable alternative duties.

Part 9 of the Act contains provisions granting inspectors the power to enter and examine any workplace (s 98 to 101). Inspectors have the power to issue non-disturbance, improvement or prohibition notices (s 110 to 115), and notices issued by inspectors must be complied with.

The Occupational Health and Safety Act deals with the establishment of health and safety representatives and committees, and provides representatives with the power to issue provisional improvement notices to the employer. This is discussed in detail at ¶2-140.

A number of regulations made under the previous legislation (the *Occupational Health and Safety Act 1985*) will continue for two years after the commencement of this Act, or until they are revoked. Requirements regarding the notification of incidents are now contained in s 37 and 38 of the Act. The Act also allows for the approval of compliance codes for "the purpose of providing practical guidance to persons who have duties or obligations under this Act or the regulations" (s 149). A number of codes of practice were approved under the previous Act (see ¶2-160). The codes do not have legislative force, but can be used in proceedings under the Act to help establish a breach of the Act or regulations.

Other relevant legislation includes:

- *Equipment (Public Safety) Act 1994*
- *Building Act 1993*: provides for building standards
- *Dangerous Goods Act 1985*, and
- *Health Act 1958*: Regulations made under the Act cover, among other things, confined spaces, fire prevention, harmful gases and hearing conservation.

Queensland

The *Workplace Health and Safety Act 1995* is the principal legislation in Queensland. The administering body is the Workplace Health and Safety section of the Department of Industrial Relations. The Act's objective is to ensure freedom from disease and injury, and the risk of disease and injury, caused by workplaces, workplace activities or specified high-risk plant (s 7).

Obligations for ensuring health and safety fall on employers, self-employed persons, persons in control of workplaces, principal contractors, designers, manufacturers, importers and suppliers of substances, and owners of high-risk plant (s 23). Each of these must ensure the health and safety of both themselves and other people in the workplace, that plant and substances used at the workplace are safe, and that risks are minimised (s 28–36).

The employer, self-employed person or principal contractor must record any injuries, illnesses or dangerous events which occur (Workplace Health and Safety (Miscellaneous) Regulation 1995 (s 15)).

Inspectors appointed for the purposes of the Act are given the power to enter any premises and carry out such examinations as are necessary to ensure compliance with the Act (s 104). Where an inspector is of the opinion that a person is contravening the Act, the inspector may issue an improvement notice, and failure to comply with such a notice is an offence under the Act (s 117). In more extreme circumstances, the inspector may issue a prohibition notice, requiring certain work or the use of a certain substance to stop until the safety risk is remedied (s 118).

The Workplace Health and Safety Act (s 92) requires the appointment by the employer of a "qualified person" as health and safety officer in all workplaces with 30 or more workers. To become suitably qualified, a person must apply for a certificate of authority to the chief executive of the Workplace Health and Safety Council. The health and safety officer tells the employer or principal contractor about the overall state of health and safety in the workplace; conducts inspections to identify hazards or unsafe conditions and practices and reports these to the employer; reports any work-caused illness, injury or dangerous event to the employer; establishes educational programs in workplace health and safety; investigates or assists in the investigation of work-caused illnesses and dangerous events; and assists inspectors with their duties (s 96).

The Act also allows for the establishment of health and safety representatives and committees (see ¶2-140).

Regulations made pursuant to the Act cover a wide range of matters. In addition, the Act provides for the creation of compliance standards (which are subordinate legislation) to minimise exposure to risk (s 38), and advisory standards (s 40) to identify and manage exposure to risk at the workplace. Advisory standards do not have the force of law, but can be used as evidence in proceedings related to alleged contravention of the Act (s 42).

Other legislation relevant to health and safety includes:

- *Building Act 1975*: prescribes standards relating to building construction, including fire safety requirements
- *Mining and Quarrying Safety and Health Act 1999*, and
- *Explosives Act 1999*.

South Australia

The principal legislation in South Australia is the *Occupational Health, Safety and Welfare Act 1986*. It is administered by the WorkCover Corporation of South Australia. The Act establishes the Occupational Health, Safety and Welfare Advisory Committee (s 7), a committee of nine members advising the government on OHS policy, administering and enforcing standards, regulations and codes of practice, and advising on their implementation and revision, among other things (s 8).

The Act places a general duty on employers to ensure the safety of employees, in particular to provide a safe working environment, a safe system of work, and safe plant and substances (s 19). Further, employers are required to prepare and maintain, in consultation with employees and their representatives, policies relating to health and safety in the workplace (s 20). Written statements setting out the practices and procedures adopted to protect the safety of the employees must be prepared and brought to the attention of the employees.

The Act also places general duties of care on employees (s 21), employers in relation to persons other than employees at the workplace (s 22), occupiers (s 23), designers and owners of buildings (s 23A), and manufacturers and suppliers of plant and substances (s 24).

Corporations must appoint a responsible officer to ensure that the corporation complies with its obligations under the Act (s 61). The responsible officer must reside in the state and must be the chief executive officer or a member of the governing body or, where that is not possible, a senior executive officer or officer of the corporation.

The Act provides for the election of health and safety representatives and, in certain circumstances, for the establishment of health and safety committees (see ¶2-140).

The power of an inspector to enter and examine any workplace is set out in s 38 of the Act. Inspectors are given the power to issue improvement notices (s 39) or, where the inspector is of the opinion that there is an immediate risk to health and safety, a prohibition notice (s 40). Failure to comply with either type of notice is an offence and the Act provides harsh penalties.

Regulations made under the Act cover a large number of specific activities and industries, including hazard identification and risk assessment, hazardous substances and plant safety. The Act also makes provision for codes of practice (s 63), which can be used in proceedings under the Act to establish that the employer did not exercise the required standard of care. A number of such codes presently exist (see ¶2-160).

Other relevant legislation includes:

- *Dangerous Substances Act 1979*, and
- *Radiation Protection and Control Act 1982.*

Western Australia

The principal legislation in Western Australia is the *Occupational Safety and Health Act 1984*, which establishes the WorkSafe Western Australia Commission. The Act requires employers to provide and maintain a working environment in which employees are not exposed to hazards (s 19). This includes the provision of safe workplaces, safe plant and safe systems of work, providing information and training to enable employees to carry out their work safely, and consulting and cooperating with health and safety representatives. Employers are required to notify the commission whenever there is an accident at the workplace which results in death or certain serious injuries as prescribed in the regulations (s 19(3)).

Employees are required to take reasonable care of their own safety and the safety of others at the workplace (s 20). General duties of care are also placed on employers and self-employed persons in relation to persons at the workplace other than employees (s 21), and on manufacturers and suppliers of plant (s 23). Manufacturers and suppliers of substances are required to ensure that adequate toxicological data are provided with the substance (s 23(3)).

The Act provides for meetings between the employer and the safety and health representative(s)/committee or the employees in order to resolve safety and health disputes (s 24; see also reg 202). Where a dispute cannot be resolved, an inspector may be notified and may attend the workplace and take such action as he/she sees fit, including an order to cease work (s 25). An employee with reasonable grounds for believing that there is an imminent and serious safety threat may cease work, although he/she must immediately notify the employer and the safety and health representative (s 26). Such employees will still be entitled to their usual pay, although the employer is entitled to require them to perform reasonable alternative work.

The Act provides for the election of safety and health representatives and, in certain circumstances, for the establishment of health and safety committees (see ¶2-140).

Section 43 governs the powers of inspectors to enter and examine any workplace. Inspectors are also given the power to issue improvement notices (s 48) or prohibition notices (s 49). Such notices may include directions as to the measures an employer must take, and failure to comply with a notice is an offence. Safety and health representatives also have the power to issue provisional improvement notices for the area/workplace/group they represent.

The Act allows for the minister to approve codes of practice for "the purpose of providing practical guidance to employers, self-employed persons [and] employees" (s 57). Such codes are not given any legislative force or evidentiary weight in proceedings under the Act.

The Occupational Safety and Health Regulations 1996 cover a large range of matters, including noise protection, workplace amenities, fire safety equipment, use of plant, first aid facilities and hazardous substances.

Other relevant legislation includes:

- *Explosives and Dangerous Goods Act 1961*
- *Mines Safety and Inspection Act 1994*, and
- *Radiation Safety Act 1975.*

Tasmania

The principal legislation in Tasmania is the *Workplace Health and Safety Act 1995*, which is administered by Workplace Standards Tasmania. The Act relates the duties and obligations of employers, responsible officers, self-employed persons, designers, manufacturers, importers, suppliers and installers regarding workplace health and safety (s 9–21). Codes of practice may be approved under the Act (s 22).

Serious accidents and dangerous incidents are to be reported to an inspector by the quickest means available (s 47). Regulations to the Act require a record of work injuries to be kept by the employer.

An inspector may direct the employer to remedy or alleviate dangerous circumstances in the workplace by a notice in writing (s 38).

The Act also provides for the election of safety representatives, with further details relating to safety representatives being contained in the regulations (see ¶2-140). Regulations made pursuant to the Act cover a wide range of specific matters, including fire safety, noise protection and machine safety.

Other relevant legislation includes:

- *Dangerous Goods Act 1998*, and
- *Radiation Control Act 1977*.

Australian Capital Territory

The principal legislation in the Australian Capital Territory is the *Occupational Health and Safety Act 1989*, administered by the ACT WorkCover Authority. The Act establishes the Occupational Health and Safety Council, a nine-member council charged with promoting occupational health and safety in the territory.

Employers in the territory are obliged to take all reasonably practicable steps to protect the health, welfare and safety at work of employees (s 27) and other persons at or near the workplace (s 28). The Act also places general duties of care on persons who have control of workplaces (s 29), employees (s 30), self-employed persons (s 31), manufacturers (s 32), suppliers (s 33) and those erecting or installing plant (s 34).

The Act requires employers with 10 or more employees to establish designated work groups, the members of which elect health and safety representatives (see ¶2-140). Health and safety representatives have the power to issue improvement notices in certain situations (s 51), or to order an employee to cease work where there is an immediate threat to the health or safety of the employee (s 56).

The *Occupational Health and Safety Act 1989* also makes provision for inspectors. Inspectors have the power to issue improvement notices where they believe there is a contravention of the Act or the regulations (s 76), or prohibition notices where they believe there is a risk of serious and imminent injury (s 77).

Employers are required to give the Occupational Health and Safety Registrar notice of a death or injury at or near the workplace (s 85). Records of such injuries must be kept (s 86). Regulations and codes of practice may be made pursuant to the Act.

Other relevant legislation includes:

- *Machinery Act 1949*: Regulations made pursuant to the Act cover the installation, operation and inspection of machinery; they include the Boiler and Pressure Vessels Regulations
- *Dangerous Substances Act 2004*, and
- *Smoke-free Areas (Enclosed Public Places) Act 1994*.

Northern Territory

The *Work Health Act 1986* deals with both occupational health and safety and workers compensation in the Northern Territory. The Act is administered by NT WorkSafe, which is the responsibility of the territory's Minister for Employment, Education and Training.

The Act places a general duty on employers to provide and maintain, so far as is practicable, a workplace that is safe and without risk to the health of workers or others at the workplace (s 29). General duties of care are also placed on occupiers (s 30), self-employed persons (s 30A), manufacturers, importers, suppliers, erectors, installers (s 30B), owners (s 30C) and workers (s 31). Where there is an immediate risk of severe injury a worker may cease work in that area (s 32), although the Act allows an employer to assign such a worker to alternative work.

The Act allows the authority to appoint work health officers (s 35), who have the power to issue improvement notices (s 40) or prohibition notices (s 41). Work health officers are also given the power to carry out investigations into health and safety matters (s 37).

The Work Health (Occupational Health and Safety) Regulations 1992 deal with hazardous substances, plant, construction work, the work environment and specific work practices. Codes of practice may be approved by the minister. While such codes are not legislation, they may be used in proceedings under the Act to establish a breach of the Act.

Other legislation relevant to occupational health and safety includes:

- *Dangerous Goods Act 1998*
- *Mining Management Act 2001*, and
- *Radiation (Safety Control) Act 1978*.

¶2-140 Legislation requiring consultation with employees

Employee consultation is the process by which employers and their employees can work out their own rules and procedures — in the context of legislative requirements — for improving occupational health and safety in their workplaces. Consultation acknowledges that everyone in the workplace has a role to play in identifying and resolving the particular health and safety problems at their workplace.

Legislation in each state and territory provides for employee consultation, mainly through the establishment of OHS representatives and/or committees. This paragraph provides a summary of those requirements.

New South Wales

In New South Wales the *Occupational Health and Safety Act 2000* and the Occupational Health and Safety Regulation 2001 make provision for employee consultation through workplace health and safety committees, OHS representatives, or some other arrangement agreed between the employer and employees. The Act specifies:

- the nature of consultation (that is, the sharing of relevant information, the opportunity for employees to express their views and the taking into account of those views by the employer)
- that consultation is required when relevant decisions are made that may affect occupational health, safety and welfare
- that consultation is undertaken with an OHS committee, with an OHS representative or in accordance with other agreed arrangements (or any combination of those means)
- the functions of OHS committees and representatives
- that more detailed provisions may be prescribed by Regulation, and
- maximum penalties for a breach of the duty to consult.

The main functions of OHS committees and representatives are also specified. These include:

- keeping OHS measures under review
- investigating any matter that may be a risk to health and safety in the workplace, and
- attempting to resolve the matter, but if unable to do so, requesting an investigation by an inspector for that purpose.

Chapter 3 of the Regulation covers workplace arrangements for consultation. An important principle to consider is that OHS committees or representatives are now elected to "workgroups" rather than "workplaces". The difference between these two terms is essentially that a workgroup consists of all the employees of an employer rather than just those employed in one location. Note that the term "workgroup" has a different meaning under other legislation, for example, Victoria's.

Another important feature of the present legislation is that employers have to be proactive and initiate consultation with the employees in given circumstances, as specified in s 13 to 15 of the Act, namely *before* decisions are made. It is not sufficient for employers to merely wait until employee representatives raise safety concerns with management.

A code of practice on consultation was released in September 2001 and is available from WorkCover NSW. It provides guidelines on the options now available, helping to interpret the new provisions. It also gives guidance on establishing the new mechanisms and suitable procedures to follow.

Victoria

Part 4 of the *Occupational Health and Safety Act 2004* deals with the duty of employers to consult with employees. Section 35 sets out the circumstances where employers are required to consult employees, and s 36 specifies that employers must share relevant information with the employees, give them a reasonable opportunity to express their views about the matter in question, and take their views into account. Sections 35 and 36 will commence by 1 July 2006 (at the latest).

The Act provides for the determination of "designated workgroups". Sections 43 to 46 allow an employee to request an employer to establish "designated workgroups" at the workplace. The groupings may be formulated on bases such as the nature of the work or hazards at the workplace, or the geographic locations of workers. Each such group may then elect a health and safety representative.

Health and safety representatives are given the power to inspect the workplace at any time after giving reasonable notice to the employer, or immediately in the event of any accident or dangerous occurrence (s 58). They are also entitled to accompany an inspector during any inspection of the workplace, and to call in assistance from an expert whenever necessary. Employers are required to provide representatives with access to health and safety information and to consult with them on all proposed changes to the workplace. Representatives are entitled to time off without loss of pay to attend training courses and to carry out their duties (s 67–68).

Further, health and safety representatives are given the power to issue provisional improvement notices (s 60). It is an offence for an employer to fail to comply with such a notice, although the employer may require an inspector to attend the workplace to confirm or cancel the notice. There is also scope for representatives to consult with the employer over safety issues, and to order a cessation of work where there is an immediate health and safety threat to any employee (s 74). Sections 60 and 74 commenced on 1 July 2005.

The Victorian Act also makes provision for the establishment of health and safety committees (s 72). A representative is given the right to require the establishment of a joint health and safety committee (s 58). At least half the members of such a committee must be employees. The committee should meet at least every three months.

Queensland

The *Workplace Health and Safety Act 1995* requires employers or principal contractors to appoint a workplace health and safety officer for workplaces with 30 or more employees (s 93–94). The officer is required to tell the employer or principal contractor about the overall state of health and safety at the workplace, conduct inspections and identify and report any hazards and unsafe practices, establish educational programs where necessary, investigate work-related injuries, illnesses and dangerous events, and report such occurrences to the employer or principal contractor (s 96).

The Act allows employees to elect a health and safety representative (s 67). They may elect more than one if the employer and employees agree. The representative is entitled to inspect the workplace; to be told by the employer of any work injury, illness or dangerous event; to report any such issues to the employer; to review such circumstances and make recommendations to the employer; to be consulted by the employer on any proposed change (to the workplace, plant or substances used) that affects health and safety; and to help resolve issues regarding workplace health and safety. The representative may also ask the employer to establish a workplace health and safety committee, and may be a member of that committee (s 81).

The employer or principal contractor must establish a health and safety committee within 28 days of a request from the representative (s 86). The chief executive of the Queensland Department of Industrial Relations may direct the establishment of such a committee. The composition of the committee is left to the employer's discretion, although it must comprise at least two people and include any health and safety officers and representatives. At least half must be workers elected by the employees (s 87).

The committee's functions are assisting cooperation between the employer and employees to ensure workplace health and safety; considering appropriate training and education; informing workers about the standards and procedures affecting workplace health and safety; helping to resolve issues of health and safety; reviewing the circumstances surrounding workplace illnesses, injuries and dangerous events; and making recommendations to the employer (s 90).

South Australia

The *Occupational Health, Safety and Welfare Act 1986* provides for workgroups to elect a health and safety representative to represent the group (s 27). The composition of workgroups should be determined by consultation between the employer and the relevant employees, or a representative of them. Section 28 of the Act, along with the Occupational Health, Safety and Welfare Regulations 1995, sets out the procedure for the election of representatives. Such elections must be by secret ballot if any member of the workgroup so requests.

Representatives are able to inspect the workplace at any time after giving reasonable notice to the employer, or immediately in the event of an accident (s 32(1)). They are entitled to bring in health and safety experts during such inspections. They are also entitled to accompany an inspector during any inspection of the workplace. Further, representatives may investigate any safety complaints made by members of the workgroup, and they are entitled to be present at any interview between the employer and an employee, or the inspector and an employee, which concerns health and safety.

Employers are obliged to consult with health and safety representatives regarding any safety procedures at the workplace and any proposed changes to the workplace (s 34). They must also provide representatives with access to safety information regarding the workplace, and allow a representative time off work, without loss of pay, in order to fulfil his/her functions and to attend approved training courses.

Health and safety representatives are also given the power to issue default notices where they believe the employer is contravening a provision of the Act (s 35), and to direct an employee to cease work where they are of the opinion that there is an immediate threat to the health or safety of that employee (s 36). In either case, representatives must first consult with the employer.

The South Australian Act also makes provision for health and safety committees where the employer employs 20 or more employees. Such an employer must establish a committee if requested by a health and safety representative or at least five employees (s 31 and Occupational Health, Safety and Welfare Regulations 1995, reg 6.2.1). The composition of the committee should be determined by agreement between the employer, the health and safety representative and any interested employees. At least half of the members of the committee must be employees (excluding managerial employees).

The functions of such committees include assisting in the formulation and dissemination of health and safety procedures, and assisting in the resolution of any safety disputes. The employer is obliged to consult with the committee over health and safety procedures to be used at the workplace, and over any proposed changes to the workplace.

Western Australia

The *Occupational Safety and Health Act 1984* and the Occupational Safety and Health Regulations 1996 make provision for both safety and health representatives and committees.

Any employee may give notice to the employer requiring the election of a safety and health representative at the workplace (s 29). Where an employer is given such a notice, it must consult with employee representatives on the number of safety and health representatives to be elected, the manner in which the election will be conducted, and the training to be given to such representatives (s 30). The election is to be conducted by a person agreed upon between the employer and employees (s 31).

A safety and health representative is able to inspect the workplace at times agreed with the employer (the employer must allow inspections at least every 30 days), or immediately in the event of an accident (s 33). Other functions of the representative include reporting any safety hazards to the employer and liaising with the employer and employees on health and safety matters.

A qualified safety and health representative may also issue a provisional improvement notice about a previously raised safety and health issue in the workplace, under some circumstances.

The employer is obliged to make available to a representative any information the employer has relating to safety and health at the workplace (s 35). An employer is also obliged to confer with safety and health representatives about intended changes to the workplace, to permit representatives to be present at interviews between the employer and an employee concerning safety matters, and to notify representatives of any accidents or dangerous occurrences at the workplace. The employer must allow a representative time off, without loss of pay, to perform his/her functions and to attend accredited training courses.

The safety and health representative may request the employer to establish a safety and health committee at any workplace of 10 or more employees (s 36). The committee should be made up of the safety and health representative(s) for the workplace, persons elected by the employees and persons nominated by the employer (s 38). The person(s) nominated by the employer shall be, or shall include, a person with the authority to put into practice such matters as the committee might resolve. At least half of the members must be either safety and health representatives or persons elected by the employees. Committees must meet at least every three months (s 41).

The functions of a safety and health committee, as set out in s 40, include facilitating cooperation between the employer and the employees relating to safety and health measures, and keeping itself informed of the standards of safety and health at comparable workplaces. The committee should consider any matters that are referred to it by a safety and health representative.

Tasmania

The *Workplace Health and Safety Act 1995* and the Workplace Health and Safety Regulations 1998 provide for both employee safety representatives and committees.

The legislation covers the election of employees' representatives where 10 or more people are employed, and the establishment of health and safety committees where there are more than 20 workers and the majority of them request it.

The functions of safety representatives are to encourage safe work practices, represent the health and safety interests of employees, and bring any hazards to the employer's notice. The representative may inspect the workplace or accompany OHS inspectors, and has the right to access all information reasonably available to the employer relating to hazards at the workplace and the results of any inquiry into accidents, injuries and occupational illness at the workplace.

The number of committee members and the committee's composition are to be agreed between the employer and the majority of employees. Unless agreed otherwise, not less than half of the committee are to be employees elected by their fellow employees. The committee's overall function is to facilitate consultation and cooperation on health and safety matters between the employer and employees (s 28). It makes recommendations regarding training and promotion of health and safety, and the establishment and monitoring of appropriate programs. A person nominated by the committee may inspect the workplace on giving reasonable notice to the employer (s 29). The employer must, among other things, make information concerning workplace health and safety available to the committee, consult with the committee on proposed changes to the workplace which may affect health and safety, and notify the committee in the event of an accident or dangerous occurrence (s 31).

Australian Capital Territory

The *Occupational Health and Safety Act 1989* and the Occupational Health and Safety Regulations 1991 make provision for health and safety representatives and, to a lesser degree, for health and safety committees. The requirements of the Act relating to health and safety representatives only apply to employers who employ more than 10 employees. However, where a principal contractor has substantial control over subcontractors at a construction site, the principal contractor may apply to the Occupational Health and Safety Registrar for a declaration that these requirements apply to that construction site, with the principal contractor becoming the employer (s 39).

An employer must establish designated work groups at the workplace (s 37), with each group then selecting one health and safety representative (s 40). A representative is entitled to time off work, without loss of pay, to attend an approved training program. The employer is liable to pay the fees for the program and any expenses reasonably incurred in attending the program.

Health and safety representatives are entitled to inspect the workplace upon giving reasonable notice to the employer, or when there has been an accident, or when the representative believes that there is a threat of an accident. They are entitled to accompany an inspector on any inspection of the workplace. They may investigate complaints by employees and are entitled to be present at interviews between the employer and employees concerning health and safety. They may have access to information held by the employer regarding health and safety risks at the workplace. Further, where the employer is requested to do so by a representative, he or she must consult with the representative on the implementation of any changes at the workplace which may affect the health and safety of employees.

Health and safety representatives have the power to serve a provisional improvement notice on a person where they believe that person is contravening the Act, and consultation with the person has not rectified the situation (s 51). They also have the power to direct that an employee cease work where they believe there is an immediate threat to the health and safety of that employee (s 56).

The Act makes no reference to the composition of health and safety committees, merely stating their functions. These functions include assisting the employer to develop and implement health and safety measures, facilitating cooperation between the employer and the employees relating to health and safety matters (s 58), and disseminating appropriate information (in appropriate languages).

Northern Territory

The *Work Health Act 1986* and the Work Health (Occupational Health and Safety) Regulations 1992 provide for consultation through health and safety committees. If requested by the majority of workers, the employer must establish a health and safety committee, if there are more than 20 workers in the workplace (s 44A). The committee consists of workers elected by their fellow workers, and persons appointed by the employer. The total number on the committee is agreed upon by the employer and the workers; half or more must be workers elected by their fellow workers.

The committee facilitates consultation and cooperation between the employer and employees. It must keep itself informed of standards and procedures relating to workplace health and safety, recommend to the employer appropriate programs and procedures, and recommend any necessary training or education relating to workplace health and safety (s 44C). A person nominated by the committee may inspect the workplace (s 44D). Employers are required to keep the committee informed about hazards and dangerous occurrences, consult on proposed changes to the workplace, and provide assistance to the committee in the carrying out of its functions.

¶2-150 Enforcement

Each OHS Act provides for its enforcement by governmental bodies operating through their inspectorates, with a scale of penalties for non-compliance. Inspectors are required to furnish proof of any failure to comply.

In general, inspectors have the power to ensure compliance through the issuing of improvement notices or prohibition notices. Improvement notices are issued where there is a breach of an Act or the likelihood of such a breach, while prohibition notices are reserved for more serious breaches and generally state that work in the relevant area must cease until the breach is rectified. In some states, health and safety representatives are able to play a part in the enforcement of the Act by issuing provisional improvement notices.

Inspectors can initiate prosecutions which, if they lead to convictions, can result in significant fines and, in some jurisdictions they can also issue infringement notices (on-the-spot fines). For full details, reference should be made to the legislation referred to at ¶2-130. Under New South Wales provisions there is also the power for authorised officers of industrial unions to initiate prosecutions.

In practice, prosecutions for the breach of a general duty of care usually occur when an accident has alerted the inspectorate to an unsafe workplace.

Limitation of enforcement

Although legislation may act as a framework for planning occupational health and safety, it alone is not sufficient. For example, it can require employers to assess manual handling risks but it cannot ensure that the employer will understand or implement this process. Effective prevention of work injury and disease requires a conscientious commitment by employers, as well as employees and supervisors, to improving health and safety conditions.

Information and education are essential, for larger as well as smaller establishments, where ignorance is often cited as the cause of contravening the law. Education needs to promote a sense of personal involvement and responsibility for safety and health in all those at the workplace. Together with the delegation of specific responsibility to supervising staff and the encouragement of active health and safety committees and representatives, it can promote safe work practices.

Legal defences

Occupational health and safety laws in all Australian states and territories make provision for employers to use certain defences if they are prosecuted following a work accident or injury. In New South Wales, for example, under s 28 of the *Occupational Health and Safety Act 2000*, the available defences may be of use to employers who can demonstrate that:

- it was not reasonably practicable to comply with the relevant provision of the legislation, or
- the offence was committed due to causes over which the person had no control, and against the happening of which it was impracticable for the person to make provision.

To use these defences, employers have to be able to argue that they did all that was reasonably practicable to meet their obligations. Under Queensland legislation, following relevant guidance material — such as the relevant regulation, advisory standard or code of practice — may also enable employers to establish a successful defence.

Employers in other jurisdictions are in a similar position — prosecutions may fail or convictions may be overturned by higher courts if employers can show it was not reasonably practicable for them to control the risk. "Practicable" means reasonably feasible having regard to a range of considerations, including the degree and probability of the risk, the state of knowledge about the risk and about the means of controlling it, as well as the availability, suitability and cost of those means.

> Successful health and safety management means going beyond the legislative requirements and adopting a proactive approach to managing OHS risks at the workplace.

¶2-160 Codes and standards

In recent years there has been increased publication of codes of practice dealing with some of the more dangerous industrial areas, as well as particular issues and procedures. These codes are intended to be used in addition to Acts and regulations.

Codes of practice are not legislation and cannot be enforced in the same manner as legislation. However, they do have the backing of the governmental body by whom they are published. A code of practice becomes effective when approved according to an occupational health and safety statute. An approved code of practice is generally designed to be used in conjunction with the statute and regulations, but a person or company cannot be prosecuted just for failing to comply with it. In proceedings for contravention or failure to comply with the occupational health and safety legislation, failure to observe an approved code of practice may be admissible as evidence. Codes of practice are of particular value to managers, supervisors, contractors and others because of the detail they provide on safe practice in specific areas.

The occupational health and safety legislation in all states and territories makes reference to the approval of codes of practice in specific areas.

In addition to the codes published by state bodies, there are National Codes of Practice published by the National Occupational Health and Safety Commission (NOHSC). These codes have no legislative force unless they are adopted into legislation by state agencies, although they can be considered as authoritative guides to good practice in the areas that they cover. The National Codes of Practice have, in some cases, been used as the basis for corresponding state codes. Continuation of this trend will lead to a more uniform national approach to occupational health and safety.

Standards

Like codes of practice, standards are not (of themselves) legislation but are accepted as authoritative guides to good practice. Australian standards have in many instances been incorporated into regulations. Standards tend to deal with specific pieces of equipment or industrial processes.

Australian standards are published by Standards Australia, a non-governmental body (see Appendix A). Each standard is designated by a number, so that standards can be referred to as AS XXXX-YEAR (for example, AS/NZS 3760-2001 is a joint Australian/New Zealand Standard entitled *In-service inspection and testing of electrical equipment*). There are thousands of Australian standards, and catalogues can be obtained from Standards Australia or viewed on the organisation's website at www.standards.com.au. It is worthwhile for employers to obtain copies of Australian standards relevant to their workplace.

The National Occupational Health and Safety Commission developed a number of standards relevant to occupational health and safety. Many of these have been adopted by the states and territories as codes of practice or as the basis for legislative requirements. For example, the *National Standard and Code of Practice for Manual Handling* is the basis for legislative provisions on manual handling in all jurisdictions. In February 2005 the NOHSC was replaced by the Australian Safety and Compensation Council, which will continue NOHSC's role in relation to the development of national standards and codes of practice.

¶2-170 Common law liability

Under common law every employer has a "duty of care" to employees and others. This means employers should provide:

(a) reasonably competent staff

(b) a sufficient number of workers to carry out the work safely

(c) a place to work that is safe and without risks to health

(d) proper plant and equipment, and

(e) a safe system or method of work.

An employee has the right to claim damages from the employer for injury or sickness incurred in the course of employment if the injury is due to negligence on the part of the employer or its agents (such as supervisors — see discussion on vicarious liability below). Negligence can occur through acts or omissions. Failure to comply with the duties mentioned above can amount to negligence.

Common law is based upon the decisions laid down by previous cases, which have built up a substantial body of law. Unlike workers compensation it considers the concept of "fault", and amounts awarded for damages under common law settlements can be substantially higher, depending upon degree of fault, extent of pain and suffering, age, disfigurement, and expected future earnings.

Note that common law and workers compensation cancel each other out; that is, an employee cannot obtain settlements under both for the same injury. Any common law settlement is reduced by the amount of any previous workers compensation settlement where both are applied for. In some states the workers compensation legislation abolishes or restricts the right to common law actions. Common law actions by injured employees have been abolished in the Northern Territory and South Australia, are restricted in New South Wales and Western Australia, and are concurrent in Queensland and Tasmania. In Victoria, common law actions were abolished in 1997 and restored in October 1999.

To defend itself successfully against such a claim the employer must prove that the measures taken to prevent such injuries, including those taken by other employees acting as the employer's agents, are in accordance with current good practice and that all reasonable care has been taken. As the techniques of health care and accident prevention have improved and become more widely known, the standard of "reasonable care" has risen.

The fact that conditions at a site were not in accordance with statutory requirements and/or relevant standards can be enough to establish the liability of an employer.

In certain circumstances, contributory negligence on the part of an injured employee can be grounds for reducing the damages claimed (see the section on contributory negligence below).

The duty of care owed to employees may extend to independent contractors where an entrepreneur engages independent contractors to do work that might easily be done by employees.

Vicarious liability

An employer is responsible for the actions of employees in the course of employment in relation to injuries caused to other employees or third parties.

An act will be within the course of employment if it is:

- authorised by the employer and is a lawful act
- authorised by the employer and is an unlawful act. If the employer directs an employee to perform duties in a manner which involves the commission of an unlawful act (which includes, for the purpose of the present discussion, breach of contract and illegal acts), the employer will be liable in damages to compensate any third party suffering loss or injury, or
- an unlawful method of performing what is otherwise an authorised act.

In these cases the employer's responsibility extends to the actions of an employee performed within the scope or course of employment.

Contributory negligence

Contributory negligence is a factor at common law, and a percentage figure is determined by the court which will be deducted from the assessed damages in all states except in a particular instance in New South Wales, where it is not a ground for defence when the action is based on a breach of statute.

A comparison is generally made with what a "reasonable" person would have done, and the degree of variance is determined in relation to whether the person was instructed clearly and specifically, together with other relevant factors.

In this regard, if a person knew of an unsafe condition, yet failed to report it or take direct action to correct it, and during the course of events was injured as a result of it, then it is likely that some reduction in damages would occur to allow for the employee's contribution to his or her injury. Each employee is in effect a health and safety officer, required to report or correct unsafe conditions.

Occupier's liability

Occupier's liability refers to the duty of care owed by an occupier of premises to the world at large. Thus, it extends beyond employees to anyone else who comes to the workplace. For example, a client or a salesperson who falls and is injured because of an unmarked trench on the premises may succeed in a common law action against the occupier.

¶2-180 Workers compensation

Workers compensation is a statutory obligation placed on all employers. It aims to provide income and rehabilitation for an employee who is injured and is unable to carry on his or her usual work. Workers compensation also provides some lump sum payments in the case of loss or restriction of bodily functions. The purpose of workers compensation legislation is not to apportion blame, and there is no need to prove fault or negligence in order to succeed.

There is workers compensation legislation in each state and territory. These Acts, and the bodies that administer them, are set out in the table opposite.

Premiums are calculated on the basis of wages and salaries paid and are adjusted from year to year on the claims experience of previous years. The premium rates vary from industry to industry according to the risk involved. For example, the rate for a clerk is only a fraction of that for a tunneller.

The courts have held in many cases that people doing business as independent contractors are to be regarded as employees for the purpose of workers compensation. Accordingly, some suppliers of delivery services and some people engaged in the building industry should be regarded as employees when arranging insurance. Advice on this matter should be available from the local state/territory workers compensation authority, or from insurance companies.

See Chapter 8 for more details about workers compensation.

Injury management

Workers compensation legislation is now placing increased emphasis on the management of work injuries and occupational rehabilitation. Rehabilitation of injured employees, for example, through a graduated return to normal duties, is beneficial to the injured person and facilitates their recovery. In addition, it reduces the time for which compensation payments must be made, thus reducing the cost of the claim.

Workers compensation legislation contains requirements that either oblige or encourage employers to rehabilitate their employees. These requirements are noted in Chapter 8, in the context of the general discussion on rehabilitation.

Workers compensation legislation and administrative bodies

Federal

Safety, Rehabilitation and Compensation Act 1988

Comcare Australia

Seafarers Rehabilitation and Compensation Act 1992

Seacare

New South Wales

Workplace Injury Management and Workers Compensation Act 1998

Workers Compensation Act 1987

Workers Compensation (Dust Diseases) Act 1942

Workers Compensation (Bush Fire, Emergency and Rescue Services) Act 1987

WorkCover Authority of New South Wales

Victoria

Accident Compensation Act 1985

Accident Compensation (WorkCover Insurance) Act 1993

Accident Compensation (Occupational Health and Safety) Act 1996

Accident Compensation (Amendment) Act 1998

Victorian WorkCover Authority

Queensland

Workers' Compensation and Rehabilitation Act 2003

WorkCover Queensland

South Australia

Workers Rehabilitation and Compensation Act 1986

South Australian WorkCover Corporation

Western Australia

Workers Compensation and Injury Management Act 1981

WorkCover Western Australia

Tasmania

Workers Rehabilitation and Compensation Act 1988

WorkCover Tasmania

Australian Capital Territory

Workers Compensation Act 1951

ACT WorkCover

Northern Territory

Work Health Act 1986

NT WorkSafe

Chapter 3

RISK MANAGEMENT

¶3-100 What is risk management?

Risk management is essentially a systematic process for considering the threats and the opportunities inherent in a situation or activity, and taking steps to achieve the best outcome. Australian Standard AS/NZS 4360:2004 *Risk management* defines the term to mean "the culture, processes and structures that are directed towards realising potential opportunities whilst managing adverse effects".

A risk management approach has been applied in many different contexts to control a variety of risks, ranging from professional liability risks to financial risks. However, its most prominent application has been in the field of occupational health and safety. In this context, the focus is on eliminating or minimising adverse effects, that is, work-related injury and illness.

Its application for this purpose is generally considered to involve a three-stage process: first, risks have to be *identified*, then *assessed*, and then they have to be *controlled*. In practice, it can be helpful to break down these three stages into smaller steps, as will be explained in more detail in this chapter.

¶3-120 "Risks" and "hazards"

In a health and safety context, "hazards" are anything with the potential to harm life, health or property and "risks" are the potential outcomes of hazards and the possibility of injury, illness, damage or loss occurring as a result of hazards. The terminology used here when referring to risk management reflects that of AS/NZS 4360:2004 *Risk management*. Much of the more recent OHS legislation refers to "hazard identification" rather than "risk identification", but in fact both hazards and risks need to be identified if they are to be eliminated or controlled.

The following table illustrates the relationship between hazards and risks.

The hazard–risk relationship		
Hazard		**Risk**
Poor design of the cabin and controls of a crane, requiring the operator to make repetitive movements in an awkward posture	→	Workers may develop back strain, neck or shoulder injuries
Toxic pesticide sprayed for weed control by local council workers	→	Workers may suffer health effects if they inhale sufficient spray or absorb enough of the chemical through their skin
Electricity on power cable of a deep fat fryer (cable insulation is cracked)	→	Worker may be electrocuted
Electric arc from welding equipment (no shielding in place)	→	Workers and others in the area may sustain eye damage from welding flash
Unguarded rollers on a machine	→	Worker's hand may be drawn in and crushed
Student misbehaviour resulting in teacher constantly shouting to be heard over the noise	→	Voice strain or damage to the vocal cords
Infected blood in syringe at hospital	→	Needlestick injury may result in infection of a staff member
Work overload and intensive surveillance of workers in a call centre	→	Work stress and excessive fatigue
Noise levels of up to 120 dB(A) emitted from a jack hammer	→	Workers and others in the area may suffer irreparable hearing damage
Heavy items to be unloaded from a truck (no lifting equipment provided)	→	Workers may injure their backs or experience other musculoskeletal disorders from lifting, heaving and carrying the items
Time-pressure with which a delivery driver must cope in order to comply with delivery schedule	→	Risk that the driver will misjudge traffic situations and have an accident because of the pressure to hurry

Risk management and legal obligations

A requirement to manage risks to workers' health and safety underpins the core OHS legislation in all Australian states and territories. As governments have moved from prescriptive legislation towards performance-based provisions, undertaking the risk management process has become a principal requirement for all workplaces. Previously, employers were expected to comply with stringent rules and regulations, which specified exactly what they could and could not do with regard to workplace safety. This was the traditional "minimum standards compliance" approach referred to in Chapter 1. Since risk management has become the dominant approach, however, employers have increasingly been given the power (and the duty) to identify their own workplace problems and to find effective solutions to them.

¶3-130 Factors causing work injury and illness

Managing workplace health and safety risks requires a realistic understanding of the causes of adverse OHS outcomes. Simplistic explanations which seek to attribute injuries to workers' carelessness or to "freak" combinations of events fail to appreciate the many factors that make incidents and injuries predictable, such as lack of training, poor maintenance or the absence of safe systems of work. The contribution of factors such as these mean that accidents, as well as injuries and work-related illness, are ultimately controllable. There is rarely, if ever, a single cause of injury or disease at work. Usually many factors are involved that interact in complex ways. The contribution of each factor to the outcome varies considerably.

By following a risk management approach, each individual factor contributing to workplace injury or illness can be identified and its importance as a contributory factor assessed. The process then assists in finding appropriate means of controlling those factors that are identified as important in order to prevent work-related injury and illness.

See also ¶7-420 for more detail on causes of accidents.

¶3-140 Elements of the risk management process

The three stages of risk management referred to above (identification, assessment and control) can be achieved through the following process, as set out in AS 4360:

1. communicate and consult
2. establish the context
3. identify the risks
4. analyse the risks (the first component of risk assessment)
5. evaluate the risks (the second component of risk assessment)
6. control the risks, and
7. monitor, review and document the process.

Communication and consultation

Communication and consultation between the employer and employees on each of the steps is an essential basis of the risk management process. Employers and employees, together with health and safety representatives and committees, are in the best position to understand the risks posed by their own work practices, and find the safest way of undertaking their work activities.

Establishing the context

It is important to take into account the organisational and industrial environment when embarking on the risk management process.

Establishing the risk management context includes:

- defining the scope of the specific activity whose risks the process is intended to manage
- setting an overall time frame for completion of the various steps in the process (though the process itself should be ongoing)
- identifying the resources required and distributing the responsibilities for conducting the remainder of the process
- developing the risk evaluation criteria — in the broader risk management context, these criteria may be legal, social or financial, or may relate to industry best practice. In relation to health and safety, however, the law sets out the basic criterion — foreseeable work injuries and illness must be prevented, so far as it is reasonably practicable to do so, and
- planning the structure of the risk management process into logical elements.

The strategic context of the process is defined by:

- the OHS policy, and
- the risk management plan.

OHS policy

An OHS policy is a written document that clearly indicates the organisation's health and safety objectives and the arrangements for achieving those objectives. It should be prepared in consultation with all staff from senior management to the shop floor and must have the endorsement of the CEO. It should also detail responsibilities for health and safety at the various levels throughout the organisation. (See ¶5-100 for more discussion on OHS policies.)

Risk management plan

The risk management plan delineates the responsibilities for carrying out each stage of the risk management process, sets a timetable for implementing the process, establishes the criteria for evaluating risks and sets performance indicators to monitor and review the effectiveness of the risk management process.

Identifying risks

This is a critical step in the risk management process. A risk that is not identified will not be controlled. Accordingly, it is crucial that this step is as comprehensive as possible.

Risk identification involves the systematic investigation of all potential risks and identifying and recording the hazards that may be causing them. In simple terms, it means identifying all the possible ways in which people may be harmed. To properly undertake risk identification it is important to understand the nature of hazards, the sources of hazards, the forms in which hazards may arise and which hazards have the potential to harm life, health and property.

Hazards may be present in the workplace environment, equipment, substances and work systems. They include:

- physical hazards (eg dangerous or poorly designed machinery, excessive noise, vehicles, heat, cold, vibration, electricity, etc)
- chemical hazards (eg poisonous liquids, vapours or fumes)
- ergonomic hazards (eg tasks which involve prolonged stooping or crouching, repetitive movements or other awkward postures or movements)
- radiation hazards (eg exposure to radiation from radioactive materials or environments)
- psychological hazards (eg stress, time pressure or bullying), and
- biological hazards (eg infectious diseases, moulds, etc).

Clearly, some types of risks overlap several categories. For example, asbestos dust is considered to be a hazardous substance and so might be considered under the heading of chemical hazards. But the damage caused by asbestos has a physical basis due to the very hard, virtually indestructible asbestos fibres causing physical scarring when they are lodged in the lungs.

Similarly, if a delivery driver with a very demanding time frame within which deliveries must be made has a car accident, is it a physical hazard (a collision with another vehicle) or a psychological hazard (the pressure created by the company's guarantee that it will deliver within 24 hours)?

Other factors may also be hazardous in effect, though they are not easy to accommodate within the traditional groupings of hazards. For example, an organisational culture which gives safety a lower priority than cost, schedule and (short-term) corporate reputation is a factor in the causation of some health and safety problems.

The overriding consideration is whether there is some way a worker could be hurt by the work practices involved in the job.

It is necessary to identify all areas in the workplace and work activities where potential hazards occur and identify the risks associated with them. Note that risks outside the workplace may also be part of the job, eg drivers and couriers are still "on the job" while out on the open road. All risks entailed in the work should be considered, not only those on the employer's premises.

Identifying the range of risks inherent in the work involves activities such as:

- checking records of injuries, accidents and incidents to identify where accidents are most likely to occur
- referring to resources such as industry standards and codes of practice that may help to identify risks which occur in your industry sector
- conducting safety audits
- establishing a regime of daily, weekly, monthly and annual safety checks and inspections, and
- consulting with employees to find out where they think risks may exist.

More details on strategies for identifying hazards and risks are given in Chapter 6.

Once all risks have been identified, they must be assessed. This is a two-pronged process of analysis and evaluation. It involves considering how serious the potential harm could be and how likely it is that harm will eventuate, and deciding whether the risk is acceptable. The second stage of the process involves ranking risks in order to allocate a budget that deals with the most serious risks first.

Analysing risks

Risk analysis is the process of identifying the likelihood and consequences of an event — that is, working out how much risk is involved — taking into account existing controls.

To analyse the **likelihood** or probability of a hazard causing harm, it is necessary to consider the causes of the hazard, how the hazard might result in harm, and the different possible outcomes the hazard might produce.

Severity may be rated as "major injury" if the potential result is permanent disability for the worker, or "minor" if the consequences would be relatively trivial. In some cases, a numerical value is placed on the likelihood and severity — for example, "highly likely" may be the second highest option for the likelihood and rated as 8 on a scale from 1 to 10.

Various methods are used to determine the severity and likelihood of risks resulting from hazards. The method chosen will depend to some extent on personal preference, but also on the relative complexity of workplace operations. Terminology should be used that is easily understood by all or most people in the workplace. The progress and results of the risk management program can thus be conveyed to all involved in workplace operations.

This process of analysing risks for their likelihood and potential severity must consider the ways in which people's health and safety might be affected and should take into account factors such as the:

- nature of operations carried out
- substances used or generated
- way work is organised
- layout and condition of plant, equipment and the work environment
- training and knowledge needed to work safely, and
- control measures currently in use.

Evaluating risks

Risk evaluation is the process of deciding whether a risk is acceptable. In health and safety, a foreseeable risk of serious injury or illness is never acceptable. A risk could only be considered acceptable if the possible adverse outcome is minor. The final stage of risk evaluation involves deciding on priorities for controlling that risk.

Considering the likelihood and severity of adverse consequences, risks can be placed in one of three categories:

1. *Low risk*	—	This risk is considered trivial. Accordingly, no further action is necessary.
2. *Medium risk*	—	This risk is considered unlikely or highly unlikely, and the adverse consequences would not be very serious. Existing controls are adequate.
3. *High risk*	—	This is an unacceptable level of risk. Control measures must be developed and implemented immediately.

This information is then used to rank the risks of injury or disease so that resources can be made available to deal with the highest priority risks first.

These two processes, risk analysis and risk evaluation, together constitute risk assessment.

All of the risks associated with each hazard identified must be assessed in this way, determining whether the level of risk is acceptable and, if not, deciding on its priority for control.

In short, when assessing the risks, three essential steps must be taken:

1. Consider the probability or likelihood of an adverse outcome occurring (that is, how likely it is that the hazard will result in injury or illness).
2. Calculate or estimate the severity of the potential consequences, and rank the risks accordingly.
3. Based on these two factors, assign priority to the risk for risk control (that is, decide whether a risk should be dealt with urgently, or in the medium term, or in the longer term).

For example, a construction worker may be "highly likely" to suffer back injury from the repeated manual handling of a 45 kg load. The injury may be serious, so the priority for control would be high.

Stakeholders and key personnel should be involved in risk assessment, and the process should be documented, that is, written records should be kept of the decisions made. The conclusions reached about risks and any supporting information on how decisions were made should be included in the records. (See ¶6-200 for more details on how to undertake risk assessment.)

Controlling risks

Risk control (sometimes referred to as risk treatment) is the process of:

- identifying a range of options for controlling risks
- assessing their respective efficiency and effectiveness
- preparing risk control plans, and
- implementing risk control plans.

Identifying risk control options

Since risk is a function of both likelihood and consequences, we can control risk either by reducing the likelihood of the event or by minimising its potential consequences. An awareness of the legislation and standards should be applied to develop a range of options to control specific risks.

Risk control must be achieved using a hierarchy of controls. The primary aim of risk control is to eliminate the risk, the best way of achieving this being to eliminate the hazard. If this is not possible, the risk must be minimised using one or more of the other control options from the hierarchy. *The risk control measure selected must be the highest possible option in the hierarchy.*

The hierarchy of controls

1. Eliminate the hazard.
2. Substitute with a lesser hazard.
3. Modify the work system or process.
4. Isolate the hazard.
5. Use engineering controls.
6. Use administrative or back-up controls such as personal protective equipment.

In many cases, it will be necessary to use more than one control method. Back-up controls, such as personal protective equipment, should only be used as a last resort or as a support to other control measures.

Examples of risk control strategies

Elimination

Eliminating the hazard is clearly the most effective method of controlling the risk. This can be done wherever an alternative way of doing the job or task can be devised.

For example, one of the tasks encountered by many plumbers, electricians, fire service installers and data/communications installers consists of drilling into concrete ceilings to attach fixtures such as hooks for suspended ceilings, or brackets to secure water pipes, electricity cables or data and communications cables.

This task presents the risk of shoulder, neck and back strain due to the prolonged arching backwards of the neck and back, as well as prolonged elevation of the arms when drilling. The weight of the drill is another factor that can add to the risk of pain or injury, as is the need to work on a ladder. A large number of falls in the building/construction industry are associated with working on ladders.

Innovative methods have been devised to eliminate the need to drill into concrete ceilings. One method requires pre-placement of metal anchors before the cement is poured, and another uses extendable tools and improved adhesives to place anchors after the concrete is set. Both of these methods eliminate the need to drill holes into concrete ceilings, and have additional benefits in that the job takes less time to complete, as well as generating less dust and less noise.

Substitution

The replacement of a hazardous process or material with something less dangerous has been widely used in the context of chemicals or other hazardous substances.

For example, asbestos was highly valued in the past due to its remarkable properties — it is light, strong, flexible and extremely resistant to destruction by heat or mechanical forces. As the world has awakened to the health problems caused by asbestos, however, its use has been phased out and substituted with safer materials such as mineral wool, rock wool, glass fibre and cellulose.

The same strategy can be used with hazardous substances such as pesticides, paints and other substances with toxic ingredients. The effort to select the safest material which will do the job has been observed in many types of businesses that use chemicals, such as dry cleaning, photography and horticulture.

Modification of the work system or process

There are many instances in which this strategy can be used to control health and safety risks. For example, in a dusty work environment such as a wood-working workshop or a building site, wet sweeping or wet cleaning methods for dusty work areas can be very effective in reducing levels of dust in the air (and in the lungs of the workers).

Similarly, in a graphic arts workshop, there may be a risk of inhaling vapours when a hazardous substance is being sprayed onto an object. The risk can be much reduced if the substance is painted on with a brush instead of being sprayed on, as fewer fine droplets of the substance will be floating in the air the workers are breathing. In this way, by changing the process used, the risk is minimised.

Tools can often be modified as well, for example, water spray attachments on tools such as jackpicks, rockbreakers and scrabbling picks used in quarrying or excavation work can do much to suppress the dust generated by the process.

Administrative approaches such as documenting safe work methods, providing training and ensuring sufficient instruction and supervision can also be used to modify the risks of a job or process. For example, entry into confined spaces with potentially contaminated atmospheres has been responsible for many workplace fatalities. Work in confined spaces can be done safely, however, if the safe work practices set out in relevant legislation and standards are rigorously followed, atmospheric testing and monitoring is carried out as appropriate, and workers are thoroughly trained in the procedures.

Isolate the hazard/risk

This strategy involves isolating or enclosing the problematic work process (or machine, or substance), or physically separating the workers from it. For example, soundproof housing can be built around some noisy machinery, or a process such as developing photographs can take place in a completely enclosed machine.

Sometimes it will be the worker who is isolated, eg situated in a soundproof cubicle in a very noisy environment.

Engineering controls

Engineering control of risks can take many forms. Sometimes a process can be carried out by mechanical rather than manual means, eg unloading a truck may be done with a forklift if the loads are stacked on pallets, rather than having workers manually lift each box off the truck. (This could also be considered as eliminating a manual handling task.)

There are many examples where engineering methods have been used to change the noise-generating components of machinery and equipment, so that the modified machine can do the same task more quietly. (This could also be considered as modifying a work process, demonstrating that these categories of risk control strategies sometimes overlap.)

Back-up controls

Administrative or back-up controls include a wide range of strategies. For example, when developing new areas of a mine, an important back-up control is to make sure that accurate maps are available of old, disused mine shafts and workings in the area, to avoid the risk of breaking through into old workings. Provision of information or training, having documented safe working procedures, or devising rosters to minimise fatigue are all types of administrative controls.

The use of *warning signs* is another example of an administrative or back-up control. Statutory provision of warning signs is covered by legislation (within regulations), but for reasons of standardisation all signs should also comply with various Australian standards covering size, shape, wording, colour and use. Placarding of chemical storages is covered by a National Occupational Health and Safety Commission guidance note and the Hazchem Codes (see ¶10-260). Pictorial signs have the advantage of reaching non-English-speaking employees (otherwise signs may need to be multilingual).

Four colours used in signs identify types:

RED — warning

ORANGE — caution

GREEN — safety

BLUE — information

Manufacturers of signs can supply samples and further information. Placement of signs at the workplace should be considered during the health and safety survey, and before commissioning new or renovated plant or buildings.

Protective clothing and equipment

The least effective risk control strategy involves the use of protective clothing or equipment, but in some instances other strategies are not practicable. In the case of certain industrial processes, there are legislative requirements to provide protective equipment.

Protective equipment should only be used as a solution when work hazards cannot adequately be controlled by other means. Protective equipment may also be used as a temporary measure until a risk can be controlled by corrective measures.

Where personal protective equipment is to be used, a list of employees who require protective equipment should be compiled. Items should be fitted and issued personally, signed for and dated.

Types of equipment

The following list contains the most common types of equipment.

Head protection:	safety helmets/caps/hats, hoods.
Eye protection:	safety spectacles, goggles, face shields.
Hearing protection:	ear-muffs, ear-plugs.
Body protection:	aprons (leather, cotton, PVC, rubber), safety harnesses.
Hand protection:	gloves (wrist or elbow length chrome leather, PVC, rubber, vinyl-impregnated, cotton interlock, loop pile, stainless steel mesh), various types of barrier creams.
Foot protection:	safety shoes/boots (with steel toecap, steel inner sole, ankle or knee length).
Respiratory protection:	respirators (single or twin cartridge, canister, disposable, airline types — either half or full face coverage), hoods, self-contained breathing apparatus.
Welding protection:	goggles, helmets, hand shields, screens, aprons, coats, leggings, spats, gloves.

Statute law requires "approved" types of equipment to be used in specific circumstances. It is the responsibility of the purchaser to check the suitability and compliance of items before purchasing.

All equipment should be in accordance with the standards set by Standards Australia, with their trademark stamped upon it. This is not only a matter of quality of manufacture. Certain types of equipment must be of the exact type designed to guard against particular hazards. For instance, welding goggles need to be of the exact shade and specification designed for the particular kind of welding carried out. Respirators are of different types corresponding to the gases, fumes or dusts they are designed to protect against.

It is preferable that equipment which must be used regularly, such as goggles or respirators, be issued personally to each employee, instead of being available for common use. Employees will be more willing to wear and to care for their own equipment.

Methods of provision vary. The employer may provide the equipment or subsidise its purchase by the employee. Provisions for maintenance of equipment may also vary. Employers may wish to survey the practices of other firms in their industry or locality.

Equipment not in regular use may be kept in stores for issue when required. It should be kept in clean and hygienic condition. Provision must also be made for equipment for visitors or other personnel.

When protective equipment is essential to protect a worker from risk, its use should be compulsory just as use of machine guards is obligatory.

When introducing the compulsory use of equipment, adequate preparation measures are needed, such as consultation with employees' representatives, education, demonstrations and warning signs. Equipment manufacturers will often provide demonstrations of use of the equipment.

Above all there should be the force of example. Managers and supervisors should be seen to follow the new rules consistently.

Assessing risk control options

Risk control options should be assessed by applying a knowledge of legislation and standards, and considering the respective benefits and costs, to choose the most suitable option. When selecting control measures, take into account the level of risk. For example, higher consequence risks need to have demonstrably more reliable controls than lower consequence risks.

Preparing risk control plans

Detailed plans of implementation of the viable options will include:

- the time frame for the implementation of the control plan
- a list of the required resources
- a budget for the plan
- responsibility for implementing the plan
- performance measures for evaluating the plan, and
- mechanisms for verifying implementation of the plan.

For example, if a noise problem has been identified, and a plan is being developed to control the risk of noise-induced hearing loss for the workers, it is necessary to first consider the range of control options, such as modification or enclosure of noisy machinery, reducing reverberations with the use of sound-insulating barriers, isolating the workers from the noise and/or requiring the use of hearing protectors (for more details about noise, see ¶10-520).

Once the most appropriate control measures have been selected, these should be documented in a plan or program that will set out the time frame (eg "these measures will be implemented within three months"), name the person (or position) responsible for taking action, and state what will be done to evaluate the effectiveness of the measures taken and the means of verifying the results (eg "in three months' time further noise measurements will be taken to verify that noise levels have been brought below 85 dB(A)").

Implementing risk control plans

Successful implementation of a risk control plan requires appropriate authority and a clear allocation of responsibilities. In addition, any training required to support risk control measures should be delivered, and its effectiveness assessed. That is, it must be clear that trainees fully understand and can cooperate with measures established to control risks.

When risk control measures have been put in place, it is vital to check how successful they are in eliminating or reducing the risk. Constant monitoring and review are required.

Reviewing the process

The risk management process needs to be an ongoing one directed to constant improvement. Risks and controls have to be continuously monitored for their effectiveness. There may be internal or external factors that affect the likelihood or consequences of an adverse event, thereby augmenting the risk. This is why the risk management process must be regularly reviewed and constantly repeated, and steps must be taken to redress any flaws. This is to ensure that new risks and those overlooked in the original exercise are identified and controlled.

The monitoring and review process involves:

- systematically checking existing risk control measures to assess their effectiveness
- considering the OHS implications of any proposed changes to the workplace, and whether the effectiveness of existing controls could be compromised
- collecting data on any new risks that have arisen, and
- formulating new control measures.

Documenting the process

The risk control process should be fully documented and these records kept with other relevant risk management records. Risk control records should be attached to, cross-referenced with or otherwise filed with related:

- hazard identification records, and
- risk assessment records.

Comments may be added to the risk control record to assist in later evaluation and verification of the control's effectiveness. Particular details which should be noted include:

- any hazards arising or exacerbated as a result of the control
- difficulties in working with the control (such as awkwardness or operator discomfort)
- environmental factors impacting on the effectiveness of the control
- additional training, information, instruction or supervision needs
- work process factors impacting on the control
- periodic issues (such as peak production periods) impacting on the use of the control, and
- other issues of concern.

¶3-170 Factors influencing the success of risk management

Workplace culture and the industrial environment

The most fundamental factor affecting health and safety performance is what is commonly referred to as "safety culture". "Culture" here means attitudes, policies, rules, course of conduct or existing practice. In a good safety culture, health and safety issues are taken seriously and adequately addressed and resourced, rather than being given a low priority and ignored.

The role of workplace culture has been highlighted by investigations into a number of high-profile OHS disasters involving multiple fatalities in Australia and in the United States. In these cases, safety was not really taken seriously, and recommendations regarding safety were not followed up. Instead, management adopted a position which rejected bad news (that is, reports of safety problems) as "negativity" and downplayed the issues.

Many companies have become very interested in their internal organisational culture in recent years, as they have come to recognise its influence on overall business performance. Equally, health and safety managers are interested in the relationship between culture and safety performance. The strongest influence on culture within most business organisations is "the message from the top". The senior managers and directors of a company send out signals indicating what they expect of their employees, either consciously (for example, awarding bonus payments for particular achievements) or subconsciously (for example, the matters discussed in informal meetings, or even the area chosen to be the butt of jokes). For this reason, the attitudes of senior managers are vitally important in the effective management of health and safety risks.

Networking

To be effective, risk management must build on the existing networks within the workplace. The success of the risk management program depends, to a large extent, on the efficiency of the management structure, the lines of communication and the cooperation of staff and management at all levels. All staff must be given training in how to use the workplace structure to implement risk management.

Fundamental elements of risk management networking and communication include:

- establishing lines of responsibility for risk management throughout the organisation
- scheduling daily, weekly or monthly planning or review meetings from the shop floor to management levels where risk management is an agenda item
- conducting meetings of the workplace OHS committee(s)
- undertaking daily, weekly, or monthly walk-through safety inspections
- ensuring regular liaison between the OHS department (which may include the OHS manager, other human resources personnel, first aiders and a rehabilitation coordinator) and the rest of the workplace
- ensuring OHS information is distributed widely
- establishing systems for reporting and recording the existence of hazards, accidents, injuries and near misses (for more detail on incident investigation, reporting and recording, see Chapter 7), and
- ensuring effective transfer of relevant OHS information between levels of management, departments and staff.

Management structure and control

Clear allocation and definition of responsibilities help to ensure that:

- health and safety objectives are integrated into the overall organisational policy
- legal duties are clearly defined
- relevant information from the techniques employed to measure health and safety performance is collected and analysed
- individual training needs can be identified according to defined roles and responsibilities
- relationships between those with different health and safety responsibilities are clearly understood and channels of communication established and used
- safety awareness and accountability are highlighted and promoted throughout the entire organisation
- appropriate criteria can be defined for recruitment and selection of new employees, and
- if the risk assessment process identifies some immediate problems of a sufficient degree of risk, there is a clear understanding of whom the problems should be referred to for action.

Ensuring effective communication

An efficient system of communication is essential for management to retain control of operations. In this way, management can verify that its intentions are understood and followed — otherwise employees may develop their own interpretation of the management agenda. In addition, communication should also involve management listening to the views of employees and OHS experts and incorporating these views into management systems as appropriate. Effective two-way communication will generate an environment in which all members of the workplace community can contribute to health and safety, thereby helping to make high standards a commonly accepted goal.

In a successful organisation people understand their responsibilities and are competent to fulfil them. Communication of important issues throughout the workplace can be achieved through written material distributed via means such as bulletin boards, newsletters and memos. Meetings and face-to-face discussions are other important ways of channelling important information.

Involvement of all personnel

To achieve optimal success in the risk management process, efforts should be made to involve everyone in the workplace where they can contribute. Commitment to workplace health and safety as a whole improves if the role, progress and results of the risk management process are made available to everyone in the organisation. This can only be achieved through effective communication and networking at all levels in the workplace.

All employees should be consulted and kept informed about:

- schedules of inspections and safety audits
- upcoming safety, OHS committee and production meetings (including agenda items)
- resolutions of safety, OHS committee and production meetings
- lost-time injury statistics
- any purchases of safety equipment, plant or substances which will affect safety
- any changes in the workplace which will impact on OHS issues
- the hazards and associated risks involved in their individual duties, and
- the control measures implemented to control these risks.

Chapter 4

OHS MANAGEMENT SYSTEMS

¶4-100 What are OHS management systems?

An OHS management system (OHSMS) aims to support the achievement of the highest level of OHS performance through systematic elimination or reduction of workplace risks. An OHSMS is an agenda for action, a methodology for fulfilling an organisation's duty of care in OHS matters.

An OHSMS consists of a range of elements which, in combination, help to prevent work-related injury and illness. Such a system presents a model which incorporates the essentials of an effective process for managing health and safety risks. As such, it is a tool for use by management.

The use of this tool is not mandatory under health and safety legislation in Australia and, conversely, using an OHSMS does not, on its own, assure compliance with legislative requirements. It is, however, recommended as a means by which organisations can improve their management of health and safety.

Organisations of all different kinds and sizes can benefit from the adoption of a range of effective OHS management actions, and a systematic approach in this can optimise the outcomes for all parties — employers, workers and all others affected by the work.

For management, an OHSMS can enable an organisation to:

- benchmark and measure its OHS performance
- implement a process of continuous improvement, and
- gain recognition for the standards it achieves in OHS management.

Such a system can be used by an organisation wishing to develop, implement, improve or audit its activities aimed at preventing and controlling health and safety risks.

Many different OHSMSs are available. Some have been formulated by OHS authorities and other government departments; others are proprietary systems developed by OHS consultants. All of them attempt to provide a systematic approach to OHS management.

¶4-120 Australian standards

Standards Australia has produced two linked and complementary standards to provide guidance on OHSMSs:

- AS/NZS 4804: *Occupational health and safety management systems — General guidelines on principles, systems and supporting techniques*
- AS/NZS 4801: *Occupational health and safety management systems — Specification with guidance for use.*

These standards set out the best elements of health and safety management systems which are already widely used in Australia and New Zealand. They describe how these elements relate to each other, and provide guidance on how to achieve best performance in each of the elements. The standards are intended to provide organisations with the fundamentals of an effective OHSMS that can be integrated with other management requirements. The emphasis in the standards, as in the legislation, is on the development and implementation of risk control actions which eliminate hazards or at least reduce the risk.

AS/NZS 4804 is the primary standard. It is relevant to all organisations and provides guidance on how to implement, develop or improve an OHSMS.

AS/NZS 4801 aims to set auditable criteria for an OHSMS. It establishes a framework which can be used by independent auditors or by in-house personnel to conduct an audit of an organisation's OHSMS.

The standards share common management systems principles with environmental management systems standards such as AS/NZS ISO 14001 and quality systems standards like AS/NZS ISO 9001.

They are also aligned with risk management philosophies and methods as set out in AS/NZS 4360: *Risk management*.

In these standards, OHSMSs are defined as:

> " ... that part of the overall management system which includes organisational structure, planning activities, responsibilities, practices, procedures, processes and resources for developing, implementing, achieving, reviewing and maintaining the OHS policy, and so managing the OHS risks associated with the business of the organisation."

¶4-130 Elements of OHS management systems

The main elements of an OHSMS, as reflected in the standards referred to above, are structured as follows:

1. general requirements
2. OHS policy
3. planning
4. implementation
5. consultation, communication and reporting
6. documentation, document and data control
7. risk identification, assessment and control
8. measurement and evaluation, and
9. management review.

Together, these comprise a process which organisations should be able to use to reduce workplace injury and illness, and to minimise the costs associated with workplace accidents. Undertaking this process should help satisfy ethical concerns and industrial relations considerations, as well as helping to demonstrate that the organisation is exercising due diligence in fulfilling its duty of care.

Each of these elements is further broken down into its constituent elements.

In both standards, matters covered under the heading of "Planning" include planning for the identification of hazards and the assessment and control of risks. An OHSMS can be considered to set out the administrative context for the risk management process.

The "Planning" section of both documents also addresses the consideration of legal and other requirements, the setting of objectives and targets, and the formulation of OHS management plans.

Under the heading of OHS management plans, AS/NZS 4801 notes that:

> "The creation and use of a plan is a key to the successful implementation of an OHSMS. The plans should describe how the organisation's targets will be achieved, including time-scales and personnel responsible for implementing the organisation's OHS policy. This plan may be subdivided to address specific elements of the organisation's operations. The plan should include an OHS review of new activities. The plan may include, where appropriate and practical, consideration of planning, design, production and maintenance. This may be undertaken for current and new activities, products and services."

The key activity addressed under "Implementation" in both standards is the OHS risk management process, but also covered are a range of associated activities, including:

- structure and responsibility
- ensuring capability
- training and competency
- consultation, communication and reporting
- documentation
- document and data control, and
- emergency preparedness and response.

Both standards also cover the follow-up procedures required to complete the risk management process, ie measurement and evaluation, investigation and corrective action following incidents or accidents, record keeping, review and audit.

AS/NZS 4804 also includes an appendix on rehabilitation. This appendix is an adapted extract from the National Occupational Health and Safety Commission's *Guidance Note for Best Practice Management of Occupational Injuries and Disease* [NOHSC:3021(1995)]. It outlines the philosophy, aims and principles of occupational rehabilitation.

¶4-140 Uses of OHS management systems

The standards do not provide advice on the management of particular OHS issues, but rather they describe the process an organisation can use to systematically control the level of OHS performance that it sets itself.

An organisation is free to implement the standards either with respect to the entire enterprise, or to specific units or activities of the organisation. For example, a company could choose to use AS/NZS 4801 to audit a particular area such as laboratory safety, manual handling, or workshop safety.

There may also be benefits to the organisation — and this is one of the aims of the standards — in integrating OHS with the overall management system. Based on a model of continuous improvement, use of an OHSMS should enable an enterprise to:

- establish a suitable OHS policy
- identify, assess and control health and safety risks
- identify relevant legislative requirements
- identify inadequacies in its OHS procedures and practices which may lead to non-compliance with legislation
- decide on remedial action to address any inadequacies
- set priorities, objectives and targets
- establish procedures to implement and review its OHS activities, and
- adapt to changing circumstances.

¶4-150 Checklist — OHS management systems

The following checklist may be used to survey an organisation's OHSMS in order to establish where there may be gaps or areas where performance could be improved. The checklist provides guidelines only. When designing a checklist for a particular workplace, individual circumstances, the organisation's cultural and industrial environment and relevant legislation must be taken into account. It must be remembered, of course, that a "paper system" which does not translate into effective action "on the ground" is of no real use, and will only provide a false sense of security.

To facilitate its use, the checklist is set out in questionnaire form.

Checklist — the OHS management system survey

A. Management responsibility and staffing

1. Is there a written OHS policy (see ¶5-100) signed by the most senior person? Is it publicised throughout the organisation? Is it disseminated to contractors/suppliers? Are the policy contents adequate? How often is the policy reviewed?
2. Does top management participate constructively in the health and safety function and assist in its administration? Do senior managers set a good personal example?
3. Does the workplace culture give health and safety a high priority?
4. Are persons engaged in the following roles:
   (a) OHS officers/managers — full/part-time
   (b) OHS representatives?

To whom do these persons report?

Are their duties, responsibilities and authority clearly set out?

Are the locations of these people suitable?

Do they have adequate equipment, time and facilities?

What training is provided for them?

How effective is that training?

5. Are there health and safety committees?

If so, how are they staffed (management only or joint employee/trade union/management)?

What matters are included/excluded from their attention?

To whom do they report? How much influence do they have?

How often do they meet?

What has been achieved by them?

What training is provided for committee members?

How effective is that training?

Where legislation exists, are the relevant provisions complied with?

If there is no committee in operation, what forms of communication on health and safety matters exist between employees and management?

6. How often (and to what extent) is health and safety discussed at board and management meetings?

7. Are written operating rules or safe working procedures used to control risks?

If so, how are they communicated to employees?

How often are they reviewed?

If some employees are not proficient in English, are written rules or procedures also issued in other languages or is the information conveyed by non-written means, as appropriate?

Is workers' understanding of the procedures checked?

8. How and to what extent is responsibility for health and safety delegated to:

(a) line managers

(b) supervisors, or

(c) rank-and-file employees?

Are there any problems caused by the approach(es) taken?

9. Is an effective contractor management system in place, and does it ensure proper risk management in relation to work done by contractors?

B. Accountability for health and safety

1. How does management hold all levels of managers and supervisors accountable for risk management?

2. Is there a system of costing accidents/work-related illness and compensation claims (for example, by section or division)?
3. Do performance appraisals of supervisors include discussion of their section's health and safety performance?
4. Are supervisors responsible for regular equipment, facilities and machinery inspections, accident investigations, and health and safety training sessions? What means are there of ensuring that supervisors carry out these duties?
5. What type of system operates to encourage suggestions and comments from staff and operatives (for example, a representative from each section on a safety committee)?
6. Is the OHSMS regularly reviewed by senior management to check its adequacy and effectiveness?

C. Inspecting the workplace

1. What inspections are made of the workplace layout, conditions, design and maintenance of machinery/equipment and facilities? How is compliance with health and safety rules and procedures enforced?
2. What systems are used for risk identification:
   (a) direct observation
   (b) review of relevant OHS guidance material
   (c) critical incidents
   (d) sampling techniques
   (e) statistical reports
   (f) interviews/questionnaires, and/or
   (g) others?

   Is there a need for further training in these techniques?

   How is each job broken down into component elements and movements to determine the location of any risks? How useful are the records kept from this analysis?
3. Are risk assessment and control records kept and, if so, have they been appropriately followed up and monitored?
4. Who is responsible for the inspections? To whom are the results reported?
5. What type of follow-up action is taken? By whom?
6. What procedure is followed to ensure safety of new plant, processes, operations or materials used? Is health and safety considered when purchase or installation is contemplated? (The success of initiatives here will often depend on gaining the cooperation of suppliers. It may be useful for organisations engaged in similar areas to formulate common health and safety oriented guidelines to assist suppliers.)
7. When corrective action is found to be necessary, how is it initiated and carried out? Is there a written follow-up system?

8. The list below represents items or conditions that may be found in a workplace. Obtain details of the type of safety inspection made and its frequency, or note risks arising from these matters.

Item	Frequency of inspection	Date last inspected	Comments
Equipment maintenance			
Noise levels			
Lighting			
Ventilation			
Temperature			
Humidity			
Extraction systems			
Air-conditioning			
Fumes			
Cleanliness/hygiene			
Stairways			
Passageways			
Floor coverings			
Slip/trip hazards			
Work at heights (with potential for falls)			
Lifts (incl marking maximum load)			
Scaffolding			
Hoists and cranes (incl marking maximum load)			
Ladders			
Ropes			
Chains, hooks and slings			
Forklift trucks			
Other in-plant vehicles			
Boilers (incl certification certificates)			
Ovens/furnaces			
Machinery guards			
Power presses			
Lathes			
Portable grinders			
Pedestal grinders			
Other woodworking machinery (circular saws, routers, joiners)			
Emergency isolation of machines			

Item	Frequency of inspection	Date last inspected	Comments
Power conveyors			
X-ray equipment			
Drums and belts			
Control of hazardous substances			
Power tools			
Hand tools			
Qualifications of operators (boilers and pressure vessels, crane, forklift)			
Compressed air (air lines, vessels, valves — incl certificates)			
Electrical equipment (insulation, wires, switching, earthing)			
Emergency plans and procedures			
Safe operating procedures			
Vehicle safety procedures			
Checks on vehicle users, licence inspections			
Washrooms/facilities			
First aid kits and contents			
First aid training			
Warning signs (incl multilingual)			
Precautions for confined spaces			
Fire or explosion hazards			
Protection from dangerous materials and substances			
Issue of protective clothing and equipment			
Maintenance of protective clothing and equipment			
Materials handling, incl: (a) manual handling, lifting and gripping (b) storage (c) internal transport routes			
Floor loads			
Pits and excavations			
Radiation			
Vibration from equipment			
Provisions for people with disabilities			

Item	Frequency of inspection	Date last inspected	Comments
Non-smoking areas Office furniture Shiftwork arrangements Any other items covered by legislation/awards or peculiar to this workplace			

D. Employee selection

1. Check the relevance of recruitment application forms and recruitment interviews to health and safety issues.
2. How are references and previous employment history checked?
3. Is a pre-placement medical examination given? (Note the distinction between pre-placement and pre-employment — in the former case, the employee is to be hired and a suitable position is being sought.)
4. Does the work involve any activities for which subsequent medical examinations are required for health surveillance purposes, and are subsequent medical examinations of employees arranged? Or are employees encouraged (not compulsorily) to arrange them themselves?
5. How is the information from medical examinations recorded?
6. Are any psychological or motor ability tests given at the recruitment stage?
7. Are any physical requirements of the job contained in the job specification? How are they taken into account when recruiting?

E. Training and competence (see generally ¶6-400 et seq)

1. Do new employees receive health and safety instruction or training during induction? What form does this training take?
2. When a new employee is being trained on the job:
   (a) who does the training
   (b) how is it done
   (c) are there written job instructions, work method statements, safe operating procedures, or some form of training manual used, and
   (d) do the instructions include health and safety aspects?
3. Are checks made to ensure employees' competence to carry out their tasks safely?
4. Do all workers have appropriate certificates of competency, where these are required to carry out their duties?
5. Is any training given to current employees who are transferred to different jobs?
6. Is use made of outside training courses (such as those conducted by government departments or technical colleges)? Are these courses evaluated?

7. Is there any provision for extra training of non-English-speaking employees (such as the issue of instructions in multilingual form, or the conducting of English-speaking classes wholly or partially in work time)?
8. What methods for training supervisors are used? Is health and safety instruction included in the training? Is the training evaluated?

F. Cultivating safe attitudes, behaviour and practice

1. Which of the following approaches, if any, are used to influence workers and management to adopt and promote safe behaviours? How have the strategies been evaluated to assess their contribution towards improved health and safety?

Item	Details	Evaluation
Group meetings		
Published material		
Films/audio-visual		
Posters		
Warning signs		
Lectures by outside experts/specialists/authorities		
Publication of statistics		

2. Does the organisation provide positive recognition for good practice?
3. What methods are used to encourage workers to report hazards and near misses without fear of blame or other negative consequences?
4. Have any special campaigns been adopted in the past (such as eye protection campaigns or "family nights")? How successful were these?
5. Are such initiatives of a "one-off" kind or an integral part of an ongoing program that provides regular follow-up and reinforcement?
6. How can various jobs be made more interesting for employees (and therefore more satisfying and health and safety enhancing) by removing or reducing their monotonous or repetitive aspects?

G. Risk management

1. Has the organisation established and maintained documented procedures to identify hazards and ensure OHS risks are controlled?
2. Does the organisation have procedures to ensure it complies with legislation relevant to the control of OHS risks?
3. Does the organisation formulate OHS-related objectives and targets as part of its planning process?
4. Does the organisation maintain management plans for achieving its OHS objectives?
5. Is hazard identification carried out on an ongoing basis, and do hazard identification procedures take into account:

(a) work organisation

(b) design of equipment and the workplace, as well as work processes and materials

(c) purchasing, production, installation, handling, maintenance and disposal of materials and equipment

(d) contract labour

(e) inspection, maintenance, testing and repair

(f) existing controls?

6. Are all identified risks assessed to estimate the magnitude of risk and assign priorities for control?

7. Are all risks controlled using higher-level controls (eg elimination, engineering controls) where possible rather than personal protective equipment, according to the hierarchy of hazard controls? (See ¶3-140 et seq)

8. Are controls checked for effectiveness and suitability?

9. Is supervision adequate to ensure control measures are used effectively and safe systems of work are adhered to?

10. Are OHS risk control measures reviewed regularly?

11. Is OHS performance measured and given suitable prominence in reports?

12. Is there a system for managing all documentation and data relevant to the risk management process?

H. Protective clothing and equipment

1. List the areas and work processes identified as requiring the use of protective equipment.

2. In areas where protective equipment is used, can the risk be controlled by alternative methods, higher in the hierarchy of risk control, to make protective equipment unnecessary?

3. What approved types of equipment are provided? Are they appropriate? Is the currently used equipment now superseded?

4. What are the award and/or legislative provisions regarding protective equipment?

5. Is the equipment comfortable and convenient to use?

6. Is the equipment maintained in good condition? What arrangements are made for maintenance (such as laundering of clothing or replacement of cartridges in respirators)?

7. What arrangements are there for the supply and issue of equipment?

8. Is the equipment actually being used? What means of enforcement and supervision are available?

9. Is the equipment properly fitted for each individual (for example, are beards interfering with the proper fit of face masks)?

10. Do employees fully appreciate the function of a particular safety device? (For example, the effectiveness of hearing protectors is, in part, a function of the length of time these are worn when exposed to noise. Damage to hearing can occur with exposure to levels well below that which might be felt as "uncomfortable".)

I. Safety and health research

The quality of information available dictates the quality of a health and safety program based on that information.

1. Are the services of an outside professional organisation used? Is it contacted often? How useful are the services and advice provided?
2. List the safety and health journals and publications to which the organisation subscribes and online sources of OHS guidance used. Is the information being read, and is it being used?
3. What are the main sources of the organisation's information on health and safety matters?
4. Who carries out the research on health and safety matters (if any) on the organisation's behalf?
5. Is useful information being obtained from employers' associations, insurance companies, OHS authorities, government departments or other sources?

A regularly scheduled OHS committee meeting with both employer and employee representatives provides an opportunity for all interested parties to receive the same information at the same time in the same way. This is known as "information parity" and the experience of other countries which have operated similar schemes for some time suggests that this is valuable, if not critical, to the implementation of a successful health and safety program.

J. Emergency planning and fire safety (see generally ¶9-400 and ¶9-520)

1. Does the organisation have an emergency control plan in place?
2. Is there an emergency evacuation plan?
3. How (and how effectively) is employee instruction carried out in relation to emergency preparedness?
4. How often are fire drills or evacuation drills held?
5. Are fire wardens and/or firefighting teams appointed?
6. Are fire exits clearly marked (and not blocked or locked)?
7. Are fire alarms installed? How often are they serviced or checked?
8. Are fire extinguishers accessible, with usage instructions clearly set out? How often are they checked?
9. Are precautions taken to reduce the risk of fire outbreaks (for example, due to inflammable materials not properly stored)?
10. Are all legislative requirements complied with?
11. How often are inspections held?

K. Incident investigation (see generally ¶7-400)

1. What are the most common types of accidents or incidents with the potential for injury?
2. What are the most common types of injuries?
3. What conditions and circumstances determine which incidents will be investigated, if some are not?
4. Who carries out these investigations?
5. How are the investigations carried out? Are there set procedures?
6. What types of incident investigation reports are prepared? To whom are they submitted?
7. What follow-up action is (or can be) taken? Who takes the action? Has it been effective in preventing a recurrence of any incident?
8. Have any specific risk control programs been established? How successful are/were they?
9. Are the current report or investigation forms adequate?

L. Health and safety information, accident records and analysis

1. What records and statistics are kept (such as accident frequency and incidence rates, compensation claims, or incident reports)? By whom are they kept?
2. Who uses and/or has access to these records?
3. How often are records analysed (daily, monthly or yearly)?
4. Are risk assessment and control records kept? In what form are they kept? What forms of follow-up on them occur?
5. Is a cost analysis technique applied to the records?
6. Are time comparisons kept to provide an indication of safety/accident trends within the organisation?
7. Is there any follow-up study resulting from the statistics kept? By whom is this done?
8. Is a register of injuries kept in an appropriate form?

M. Purchasing

1. Do purchasing checklists include key OHS selection criteria so that health and safety conditions can be gradually improved? For example, is a "buy quiet" policy in place so that noise problems can be reduced over time?
2. Are item specifications developed in consultation with those who will be using the equipment/substance and maintaining the equipment?
3. Are applicable standards identified and complied with?
4. Are risk assessments carried out where necessary in developing specifications and/or introducing new equipment or substance?

5. Is information on the risks sought from the supplier/manufacturer, and where there is an unsatisfactory response, is the purchase not carried through?
6. Are modifications to equipment sought where necessary to reduce risk prior to delivery?
7. Are checks made prior to acceptance to ensure the product meets the agreed specifications?
8. Are end users provided with information and/or trained in the safe use of the equipment/substance before using it, and if necessary, are safe work procedures developed?
9. Is end user satisfaction with the equipment/substance checked?

N. Medical programs and facilities (see Appendix C)

1. Which of the following staff are employed (numbers)?
   (a) first aid attendants (in each work section)
   (b) occupational health nurses — full/part-time
   (c) occupational physicians/company doctors — full/part-time or on a contract basis, and/or
   (d) occupational hygienists.
2. What makes these people appropriately qualified?
3. How often are first aid training courses held? Is this sufficient?
4. Who is responsible for first aid supplies and facilities? Does this arrangement work effectively?
5. What is the procedure for obtaining first aid assistance?
6. Are first aid facilities suitably located?
7. Are medical treatment facilities adequate and kept in good order?
8. What emergency facilities are available if the normal first aid attendant is unavailable?
9. What facilities are available for transportation of the injured to a hospital?
10. Is a directory of physicians, hospitals, ambulances available?
11. Are any health education programs carried out by the organisation? How have these been evaluated?
12. What leisure facilities are available (such as sporting, gymnasium or recreation areas)? Are these promoted by the organisation? How often are they used?
13. Are counselling and/or referral facilities available to employees with health problems such as stress or alcohol/drug addiction? (See ¶9-650)
14. Does the organisation tend to promote a "healthy lifestyle" for its employees? In what ways? (See also 12 above.)

15. Are there any particular health risks/problems at the workplace, such as harmful substances/work processes? How well do employees working with these understand the hazard(s)? What action can be taken to minimise the effects of these?

16. Are there warning signs of problems, such as "disease patterns" (for example, a number of employees suffering from headaches, nausea, rashes, eye irritations or coughing)? Are these investigated?

17. Where health hazards exist, are there adequate means of warning employees?

18. Is health surveillance carried out where appropriate or required by legislation?

O. Miscellaneous

Any other aspects of the workplace not mentioned above that require action or attention.

Item	Comments

Chapter 5

POLICY AND PLANNING

OHS POLICY

¶5-100 Formulating an OHS policy

The logical first step in managing health and safety risks is for top management to issue a policy statement on occupational health and safety. The policy should state that the organisation accepts responsibility for the safety and health of its employees and should set out management's goals, responsibility, accountability and participation in the safety/health function. The statement should aim to enlist the support of all employees, and should express support for OHS representatives and committees. It should not create a situation where what is said cannot (or fails to) be implemented at all organisational levels, otherwise the policy will lose credibility with the workforce.

A "team" approach between management and employees in the formulation of the policy statement will provide an opportunity for workers to relate to the statement and also help develop a relationship of mutual trust.

Content of statement of policy

1. Expression of management's objectives and intentions.
2. A set of general *guidelines* to the health and safety function (not to be confused with more specific procedures).
3. Acceptance of primary responsibility for the health and safety function by top management.
4. Expression of support for health and safety representatives and/or the workplace safety committee.
5. Outline of authorities and responsibilities of all employees at all levels.

The policy must be publicised, either by distribution to employees, display on noticeboards or announcement. It is important to talk to people as well as making communications in writing. Some workers may not be attuned to attending to written communications, either because English is not their first language, or they are disadvantaged in terms of functional literacy, or they are inundated with written communications, or for some other reason. Where there are workers who do not read or speak or understand English well, it is important to provide multilingual translations or convey the information by other means, eg verbal explanation and demonstration.

Policies versus programs and procedures

A distinction needs to be drawn between the role of a policy (as a general statement of intention and objectives), a program (a set of detailed steps that explain how the policy is to be implemented), and procedures (which may include safety rules or work method statements for particular tasks). The OHS risk management program will describe the organisation's plans for addressing particular issues. For example, a program on hearing conservation may cover testing procedures, noise reduction and use of ear-muffs, and it may be supported by a set of specific instructions.

¶5-110 Sample health and safety policy

See the sample health and safety policy below. This is an example of an organisation's policy, to show what items could be included. Alterations can be made to cater for each organisation's different circumstances.

In large organisations, it may be a good idea for line managers to countersign the policy, as it emphasises local responsibility and improves the status of these managers where safety and health matters are concerned.

Sample OHS policy

It is this organisation's policy that each of its employees shall be provided with a safe and healthy place in which to work, and that our work practices will not compromise the health or safety of others present at the workplace, for example, contractors, visitors or other non-employees.

To achieve this policy, management will make every reasonable effort in the areas of hazard identification, risk assessment and control, as well as health preservation and promotion. These aspects of working conditions will be given top priority in company plans, procedures, programs and job instructions.

In conjunction with this policy, a series of programs, procedures, and rules on specific individual health and safety matters will be prepared and issued.

Health and safety at work is both an individual and shared responsibility of *all* employees. The following areas of responsibility are essential to the success of the policy:

1. *Top management*. Senior management is required to actively pursue the goals set out in the first two paragraphs of this policy through the following approaches:
   (a) devising and administering a comprehensive OHS risk management program
   (b) holding regular senior staff and supervisors' meetings to discuss health and safety performances, and
   (c) taking effective action to provide and maintain safe and healthy working conditions for all employees and others affected by the undertaking.
2. *Supervisors*. Supervisors will be held accountable for working conditions under their control and for the extent of information, instruction, training and supervision given on safety and health matters to employees. They are to provide the initiative and follow-up action to maintain this policy within their own sections.
3. *All employees*. Employees share responsibility for their own safety and that of their co-workers and others affected by work practices. The success of a safety and health program ultimately rests on the willingness of everyone to cooperate and work collectively with a "team spirit".

The workplace health and safety committee/health and safety representative shall be supported by management and employees.

Reducing work-related injury and disease, as well as related insurance costs, will permit the company to be more competitive in its industry, thus helping to safeguard jobs.

Signed...................... Date......................

(Chief Executive Officer)

PLANNING RISK MANAGEMENT

¶5-200 The planning process

To meet the overall goals of the organisation's policy, specific objectives and targets need to be established during the planning phase. This entails setting performance standards, which should cover both organisational procedures and the control of specific risks. Information arising from the planning process is therefore essential for effective implementation.

For example, in planning risk management objectives, information may be sought on current industry standards with regard to the control of specific risks in that sector. This information will be extremely useful later in indicating hazards, which should be checked for during workplace risk identification exercises. Equally, statistical information on industry accident rates may be useful in determining the priority of risks during the risk assessment phase.

Elements in successful planning

To plan effectively for risk management, employers should be familiar with:

- the legal duties to employees, contractors, members of the public and others who might be at risk as a result of the organisation's activities
- relevant industry health and safety standards, and other OHS guidance material, such as guidance notes or safety guides
- the availability of in-house health and safety advice and support
- access to relevant outside services and organisations that may be able to provide advice and assistance (such as government agencies, training consultants, emergency services and statutory authorities), and
- the principles of risk management.

This requires organisations to make sure they identify, and have access to up-to-date versions of all the legal and other requirements applicable to the organisation. It also requires the development and documentation of procedures to identify, assess and control risks to health and safety.

Contribution to overall management

The process of deciding on the safest way to undertake the organisation's work activities expands management's understanding of the advantages and disadvantages of the various options. There are always some ways of working which are safer than others. Safe working methods must be defined before they can be followed, so risk management planning activities can raise managers' awareness of issues such as:

- the benefits of developing and defining safe work procedures

- prevailing health and safety risks associated with the types of plant, substances and processes existing at the workplace
- current industry practice with regard to the control of risks
- OHS implications of introducing new processes or production methods, and
- new technology available in areas such as plant safety devices, alternative substances to currently used hazardous chemicals, personal protection and work systems (for example, automation and access control).

Prerequisites to implementation

Certain matters should be resolved during the planning process. Prior to implementing a risk management program the employer should:

- identify the need for information on health and safety issues and ensure that it is disseminated as required
- ensure that effective methods of communication are established and used to promote safety awareness at all levels, and
- ensure that arrangements are in place (and that they are properly used) to achieve effective consultation between management and employees on health and safety issues.

The people asset

Effective management of health and safety risks requires engagement of upper management and the entire organisation. The selection of competent personnel as core participants is another critical factor. The responsibilities of the health and safety manager, OHS committee, OHS representatives, risk management team leaders and others involved in specific tasks within the process must be clearly defined during the planning process. All of these people must be provided with sufficient information and training to undertake their individual roles effectively.

The steps required to achieve objectives should be planned and the plan broken down into tasks that can be allocated to identified individuals or groups. These tasks must be matched to individual competencies and must conform to overall organisational systems and planning.

For example, an overall objective to reduce strain injuries will require many supervisors and managers in the organisation to understand the individual tasks required of staff. Their abilities and capacities in terms of workload must be matched to the tasks allocated. In addition, they must be provided with sufficient information, training and support (including time and resources) to carry out their tasks efficiently.

OBJECTIVES AND TARGETS

¶5-300 Setting objectives and targets

Clearly formulated objectives give the organisation a clear vision for what is to be achieved as well as a way to evaluate its OHS performance.

The objectives will depend on the program and the workplace. What is important is that the objectives chosen are measurable, rather than vague statements. Performance objectives should be based upon the "SMARTA" principle, that is, they should be specific, measurable, achievable, relevant, given a time frame, and agreed.

For example, "a better place to work" or "cleaner air in the laboratory" are not measurable objectives. Better choices would be "a 30% reduction in labour turnover within 12 months" or "reducing the concentration of compounds X, Y and Z in the laboratory atmosphere to below the National Occupational Health and Safety Commission's published exposure limits".

Where the program is one of ongoing improvement, a time frame should be specified. In some cases improved productivity, or increased safety with no loss of productivity, can form part of the objective.

While **objectives** state what is intended to be accomplished, **targets** define the performance level within a time frame. The extent to which objectives are met is then measured by **performance indicators**.

For example, in an effort to eliminate or reduce back injuries or other strain or sprain injuries associated with lifting, moving or carrying loads, an organisation could adopt the following objective, target and performance indicator:

Objective:	Eliminate manual handling injuries.
Target:	Assess risks and introduce measures to control all manual handling risks, by [date].
Indicator:	Proportion of manual handling risks assessed and for which control measures have been implemented.

More information on performance measurement is provided in Chapter 7.

Assessing results against original objectives

The reason for setting measurable objectives is to enable the outcome of a program to be assessed, and to make any changes that are shown to be necessary.

Case study

A hazardous substances safety program

The objectives of a hazardous substances safety program were agreed to be:

- catalogue and prepare MSDSs (material safety data sheets — ¶10-240) for all chemicals and other hazardous substances at the workplace
- adequately label all chemical storages at the workplace
- ensure that showers and eye-baths are available at all locations where they may be needed, and
- reduce the atmospheric level of specified chemicals to below agreed levels.

Six months was set as the target time for achieving these objectives.

At the end of six months a survey of all chemicals and other hazardous substances in the workplace had been carried out, and all chemical storages had been labelled in accordance with the NOHSC's guidelines. Showers and eye-baths had been placed within easy access of all major chemical usage areas at the workplace.

Measurements of the specified chemicals showed that atmospheric levels had in fact been reduced to below internationally accepted exposure standards. However, MSDSs had not been compiled for some of the chemicals that were catalogued. The reason for this was a delay in getting information from certain chemical supply companies. It was decided to put pressure on the companies to supply the necessary information, and to investigate other suppliers if information was not forthcoming within two months.

¶5-350 OHS management plans

To achieve its objectives and targets, the organisation can create and use an OHS management plan. The plan is the key to the successful implementation of the OHS management system.

OHS management plans should describe the means by which the targets will be achieved. Typically, they include time frames and personnel responsible for implementing the organisation's policy.

The plan can focus on specific elements of the organisation's operations, for example, on manual handling risks, or on the design of equipment, work systems and individual jobs. Where appropriate and practical, the plan may include a consideration of production and maintenance. This may be undertaken for current or new activities, products and services.

For installations or significant modifications of work processes, the OHS management plan may address the planning of the change, the design, construction, commissioning and operation of the proposed new process. The plan should also address matters such as decommissioning of plan and disposal of substances. Procedures should be established to ensure that plans are reviewed at regular and planned intervals, and amended if necessary.

ROLES AND RESPONSIBILITIES

¶5-400 Overview

This section describes the roles and responsibilities of the various types of staff involved in an organisation's OHS functions.

The use of these types of staff will depend on the organisation's own circumstances — its size, dispersion of employees, type of work, control systems for hazardous processes, quantities of dangerous materials in use, incidence of shiftwork, working conditions, proximity to outside health and medical care, and other individual aspects relevant to its style of leadership and organisational structure.

All employees at the workplace should be made aware of the various safety and health staff and the functions of each. This information should be included as part of induction training.

¶5-420 Who is responsible for health and safety?

The final responsibility for occupational health and safety lies with the person who is ultimately in charge of the workers, such as a CEO. Particular OHS responsibilities can be delegated, and persons held accountable.

The authority delegated to carry out health and safety responsibilities must operate in the same way as other management functions. It begins at the top level with an approved written policy statement, procedures, rules and instructions that, once issued, must have some type of performance assessment system to measure compliance and personal accountability.

To carry out the health and safety function effectively, the degree of authority delegated must equal the amount of responsibility given. For example, those who are accountable for the health and safety of workers in a particular area must have the authority to redesign work processes in that area. While this is relatively straightforward for organisations that employ workers on their own premises, it is more difficult when the employer is a labour hire company whose workers are hired out to other organisations.

The courts have held that in this case the employer, that is, the labour hire company, still shares responsibility for the safety of workers who are employed at other organisations. In this situation the labour hire company as well as the "host employer" are both responsible for the workers' health and safety.

Many organisations evaluate health and safety performance as one aspect of performance when considering possible promotion opportunities. Frequently, financial loss due to accidents, injuries and damage may equal or exceed the organisation's profit for the same period. Therefore it is advisable that these losses and results be included in all relevant reports, including annual reports to shareholders.

It is important to remember that the exercise of health and safety responsibilities has a flow-on effect to other aspects of the company's overall health. For example, the growth in socially responsible investment funds means that good OHS performance may represent a potential business development opportunity (or risk).

¶5-430 Responsibilities of top management

Senior management are responsible for an organisation's strategic planning in relation to occupational health and safety. This will mean determining the resources (finances, staff, time and access to consultants) that are to be devoted to health and safety. What constitutes adequate resources may be contentious, with employees and management expressing conflicting views. Senior management will be required to make the final decision and must take responsibility for an organisation's health and safety performance.

Other top management functions relevant to health and safety include:

- determining which safety programs are to receive priority
- monitoring the outcome of an organisation's health and safety programs
- ensuring compliance with legislation, awards and standards
- endorsing the formulation of appropriate rules, procedures and methods for the workplace
- ongoing and effective dissemination of OHS information and promotion of health and safety awareness in the workplace, and
- commitment by personal example.

¶5-440 Responsibilities of line managers and supervisors

Line managers and supervisors must accept as an integral part of their duties the responsibility for implementing and administering health and safety procedures at the workplace.

Supervisors/managers will have overall management responsibility for their sections, and the basic functions of planning, organising (including staffing), leading and controlling apply to the safety issue. Responsibilities include the following:

- overall supervision of employees to ensure the health and safety of the worker, the public and the consumer
- hazard identification and assessment and control of risks
- implementation of particular safety programs
- on-the-job training and instruction
- efforts to motivate employees to comply with safe work practices, including specific directives when giving orders
- accident investigation and correct reporting
- issue of and ensuring correct use and maintenance of appropriate personal protective clothing and equipment

- submission of reports and recommendations/suggestions about hazard controls, workplace procedures, etc, to more senior management, when these issues are outside their scope of authority, and
- decision-making regarding job design, workplace layout and (possibly) recruitment, where they have this authority.

Where the organisation has full-time health and safety staff (see Appendix C), this staff will provide a support or back-up role to managers and supervisors, but will not have the responsibility of the latter transferred to them.

¶5-450 The role of the human resources department

The role of a human resources department within an organisation is one of a service/advisory/strategic function, not a line management function. Frequently, the health and safety function is administered by the human resources department, where it is often one of the "extra duties" added to that function. Where this is the case, a specific time allocation must be given to enable those duties to be carried out. Otherwise, this function will be given less time due to other pressures until it becomes a function in name only. Furthermore, it should be clearly understood that the human resources department's activities in administering the health and safety function do not lessen the responsibility of line managers or supervisors, nor do they replace the requirement for employers to consult with workers.

Where the human resources department is given the task of administering or resourcing the health and safety function, some or all of the duties set out at ¶5-460 should be performed by that department. In this situation it will be worthwhile to build up a list of outside contacts and sources of technical and health and safety information, as it is likely that this type of assistance will be required.

Human resources personnel may participate in a workplace health and safety committee.

It is also appropriate for this department to be aware of the functions of occupational health and safety staff and of training available in this area. The department may be required to coordinate training for staff.

¶5-460 The health and safety officer or manager

Where organisations employ full-time OHS staff, these are usually known as the "health and safety officer (or manager)", or sometimes the function is performed by a "safety, health and environment manager". These designations are tending to replace "safety officer" and the narrower focus that title implies.

As can be seen from the list of the person's likely duties below, the job is mainly an administrative and advisory one (that is, a staff position). The officer/manager should report directly to top management, which takes action after considering the advice received. At the same time, it is necessary for the officer/manager to possess sufficient authority to take any action required in a possible emergency. For example, if the supervisor were not readily available, and a dangerous situation arose in which an accident could occur, action must be taken to prevent injury and damage. Normally, however, the position will have no formal authority over line managers or supervisors.

The duty statement below is a sample relating to a large public sector organisation.

Duties of a health and safety officer/manager

1. Conducts inspections personally or in company with executives and supervising officers, specialist consultants, or health and safety representatives/committee members to ensure the observance of health and safety standards and for the purpose of discovering unsafe or unsatisfactory conditions and practices before personal injury occurs. The health and safety officer/manager should be able to identify and qualify/quantify risks of injury or disease from occupational hazards and determine possible control measures.
2. Reports any unsafe and unsatisfactory conditions, procedures or operations to the supervisor or executive in charge (note that authority to order cessation of work in a dangerous situation must receive serious consideration, particularly where there is legislative provision for the cessation of work deemed to be unsafe (see ¶2-130)).
3. Acts as adviser to executive and supervisory staff and employees in all matters concerning prevention of accidents, injury, hazard, disease and the promotion of health and safety. Monitors the organisation's overall safety performance in order to report this to senior management.
4. Receives and reviews all reports of accidents, incidents, injuries, potential hazards or "near misses". All such reports should reach the health and safety officer as quickly as possible. Prepares reports and recommendations to management and/or safety committees/representatives.
5. Investigates selected accidents, hazards or "near misses", as distinct from the obligatory investigation made by supervisory staff, and recommends the appropriate action to prevent a recurrence.
6. Maintains an injury record system generally in accordance with legislation and relevant standards.
7. Attends all meetings of health and safety committees, and staff meetings or conferences when health and safety matters are discussed or considered.
8. Organises health and safety training of staff and employees in conjunction with executives and supervisors. Institutes health and safety promotion campaigns to create and maintain an interest in health and safety at all levels. Conducts or arranges health and safety induction courses for new employees.
9. Identifies necessary resources and is aware of the availability of specialist support advice. Where medical or nursing staff are not immediately available on the premises, ensures that first aid facilities are adequate and maintained in satisfactory condition. Also makes arrangements with ambulance, casualty and medical services to be available as required.
10. Monitors and advises — on changes to legislation and award provisions relevant to health and safety, as well as the OHS implications of changes from all sources.

11. Reviews, monitors and updates health and safety manuals, rules, procedures, etc.
12. Encourages compliance with safe work practices through distribution of information and training resources such as websites, publications, posters, warning signs, videos, etc. This material can be either general (such as an induction handbook) or on specific topics (such as handling harmful chemicals, manual handling, etc).
13. Liaises with the organisation's insurance company or broker. This person may be involved in liaison in a workplace rehabilitation program (assisting in evaluating the effectiveness of programs, etc).
14. Stays abreast of health and safety strategies within the particular industry.
15. Carries out or organises research into safety and health matters as required by the organisation.
16. Answers enquiries from management, supervisors and employees.
17. Evaluates health and safety products and systems, and advises management on their suitability and application to the organisation.

The health and safety professional should have the training to be able to recognise most of the types of hazards that exist, and be able to learn by both studying the workplace and consulting other sources of information. Good liaison between the health and safety officer and supervisors is essential as the health and safety officer could lack the technical background of the supervisor.

The qualifications relevant for OHS officers or managers are set out in Appendix C.

¶5-470 OHS representatives and committees

As noted in Chapter 2, employee consultation is required in all Australian jurisdictions, and this is mostly done through health and safety representatives and workplace OHS committees. Through these mechanisms, employers and workers can jointly tackle OHS problems, and employees can express safety concerns to management and provide management with information which they may not otherwise obtain on potential safety problems at the workplace. Active health and safety representatives and committee members will be able to generate an interest in safety throughout the area that they represent.

Generally the legislation sets out the method of selecting a representative or committee, as well as their powers and functions.

Selection of representatives

Health and safety representatives and committees typically act for one particular group of employees. These groups will often be the various departments or geographical areas within an organisation, but they may also be chosen along other criteria such as shiftworking arrangements or commonality of the hazards faced by particular employees. In smaller organisations it may be sufficient to have only one health and safety representative.

To ensure employees' confidence in OHS representatives and committee members, and to guarantee that they remain accountable to the group, OHS representatives and committees should be elected by those they will represent (see ¶2-140).

An appointment for a period of at least two years is desirable, because of the training necessary and the period required for individuals to familiarise themselves with their role. However, there is also some advantage in a rotation of employees through the position of OHS representative, because of the experience and awareness that it creates in the employees concerned.

Involvement of unions/union delegates

An issue that arises in relation to OHS representatives and committees is whether workgroups should be determined along the lines of union membership and whether union delegates should become health and safety representatives or committee members. Clearly, the health and safety of employees is a major union concern; however, health and safety issues should not become mixed up with more general employer/employee matters.

In workplaces that have poor industrial relations records, it may be best not to determine workgroups along union lines. To do so could encourage safety issues to be thought of as an extension of industrial relations issues, and may lead to an adversarial approach in relation to the safety issues that do arise.

There is no overwhelming argument either in favour of or against having union delegates as OHS representatives and committee members. In addition to the risk of merging OHS and general industrial relations issues, where one person occupies both positions, he/she may not be able to give sufficient time and energy to either. On the other hand, a good union delegate will already have the necessary negotiation and consultation skills required to succeed as a representative or committee member. In reality, a popular and effective union delegate is likely to be elected as a representative or committee member (if he/she wishes to take on the additional workload), whilst a poorly performing delegate will not be.

Functions and powers

The crucial functions of health and safety representatives and committees are to provide a mechanism for employers to consult employees about health and safety, to convey employee safety concerns to management and to represent employees in relation to the improvement of workplace health and safety.

In order to carry out the function of health and safety representative effectively, the following powers are regarded as essential (in each state, these powers are prescribed to a greater or lesser degree (see ¶2-140)).

Rights of OHS representatives and committees

- *The right to know about health and safety at the workplace.* This means suitable training must be given to representatives and committee members. It also means access to information that the employer has relating to potential dangers at the workplace. Representatives and committee members should be allowed to keep up to date with safety developments in the relevant area (courses, subscriptions to journals, tours of similar workplaces).
- *The right to inspect the workplace.* It is important that representatives and committee members conduct regular inspections of the area they represent. A checklist may be useful. The right to investigate an accident or dangerous occurrence is also usually prescribed by the legislation.
- *The right to participate in health and safety activities.* Representatives and committee members should have input into the OHS policies adopted at the workplace. They should be informed of any new equipment or processes that are being considered. The legislation may also provide a right to be present at any interview between the employer and an employee concerning a safety matter.

In some states health and safety representatives are given the power to direct employees to cease work where the workplace is considered to be unsafe, and the right to serve the employer with a provisional improvement notice where the representative considers that there is a breach of legislative requirements (see ¶2-140). In these situations the representative should first consult with the employer in an attempt to solve the problem.

The functions of workplace health and safety committees and representatives are advisory and educational, not decision-making. Management's ultimate authority, and that of supervisors, must not be undermined by the committee. Therefore, a committee can make recommendations to persons who do have the authority to implement them.

Functions of OHS committees and representatives

These functions are:

- to enable rank and file workers to take an active part in the promotion of a healthy and safe workplace
- to give employees the understanding necessary to assess the feasibility of proposed risk control measures
- to obtain the benefit of the great store of knowledge and experience possessed by many employees regarding health and safety aspects of the work they perform by obtaining feedback from them
- to review measures taken to ensure the health and safety of persons at the workplace
- to investigate matters brought to the employer's attention which a committee member or employee considers to be unsafe or a risk to health (resolution of that may involve recourse to an inspector, appointed under the relevant Act, to inspect the workplace)
- to assist in the development of recording systems for incidents and hazardous situations and to promote among employees an understanding of such matters as accident causation by reviewing recent accidents

- to assist in the development of a safe working environment and safe systems of work
- to assist in the formulation and effective implementation of an organisation's overall occupational health and safety policy
- to monitor measures taken to ensure the proper use, maintenance and (if necessary) replacement of protective equipment, and
- to make recommendations to the employer regarding health and safety matters.

Composition of OHS committees

Ideally, the workplace health and safety committee consists of representatives of several levels of management, with at least 50% of the members being elected employee representatives. The inclusion of a senior management representative will give the committee status and authority as long as the manager has the authority to make decisions. The size of the organisation, along with the number of separate departments within the organisation, may determine the size of the committee. Note that a committee of more than eight or 10 people will tend to become unworkable. The legislation in most cases provides a guide to the composition of the committee, as well as the method of selecting the employee representatives (see ¶2-140).

¶5-475 Training

Adequate training of health and safety representatives and committee members is crucial to their successful functioning in the role. Generally, the legislation states that they must be granted time off without loss of pay to attend accredited training courses (see ¶2-140). Training courses are available through government departments, union bodies and private agencies (see the organisations listed in Appendix A).

An effective OHS representative or committee can help realise the benefits which are the aim of employee consultation but, conversely, there is the possibility that the committee will become simply a "talkfest", rather than a forum for actually generating improvements. There is also the risk that the committee will become entwined in general industrial relations issues. This can be overcome by having a strong chairperson and by setting agendas (see below).

Conducting a health and safety committee meeting

Timing of meetings should be arranged according to the organisation's own needs and workloads. They should be frequent enough to avoid an excessive number of topics having to be discussed at each meeting.

A suitable time for conducting meetings could be either one hour before lunch or one hour before finishing time, thus providing an automatic time limit on meetings.

Typical duties of committee members are set out below.

1. *Chairperson*: arrange for meeting venue and time, review previous minutes and materials for meeting, arrange program, notify all members of above items and make other arrangements as necessary.
2. *Secretary*: prepare and distribute minutes, prepare agenda, forward recommendations and deal with correspondence.
3. *Members*: attend all health and safety meetings, report all hazards and instances of unreported injuries, accidents or near misses, contribute ideas and suggestions, work safely and endeavour to influence others, and carry out other duties as delegated.

¶5-480 The first aid officer

Occupational health and safety legislation generally requires employers to provide first aid facilities in their workplaces. Industrial awards may contain a similar requirement. The scale of facilities required varies with the number of employees and the type of workplace.

Employers should ensure they are familiar with the details applying in their state or territory. This information is available from OHS authorities.

First aid facilities include:

- a properly stocked first aid box (as to contents, see below)
- a stretcher
- a shower and/or eye-bath (where chemical hazards are present), and
- a first aid room (large organisations only).

Several employees should be trained in first aid, with one of them placed in charge of first aid facilities (including stock and maintenance). Many awards prescribe a first aid allowance to be paid for performing this duty.

First aid officers should hold an appropriate (and current) certificate. First aid courses are conducted by both the Australian Red Cross and the St John Ambulance Association. The names and locations of the current first aid officers should be well publicised throughout the organisation.

Provision of first aid facilities should be regarded as a minimum requirement, even if the organisation has no other form of occupational health facilities or service.

In an organisation that has its own occupational health service, first aid facilities should still be available at appropriate locations in case of emergencies or very minor injuries that do not require the attendance of occupational health staff and to cover work times when OHS service staff are not in attendance, eg shiftwork, weekends.

Occupational health staff should supervise the first aid staff in charge of these facilities and oversee the maintenance and stocking of first aid boxes, etc. First aid staff should still be appointed from the shop, office, construction site or factory floor.

General duties of a first aid officer

1. Dispense and control items from first aid cabinet.
2. Ensure cabinet supplies are adequate.
3. Treat minor wounds and injuries, such as applying dressings, stopping bleeding and treating burns.
4. Deal with fits/fainting.
5. Resuscitation.
6. Record accident/injury details in accident book.
7. Arrange further assistance if required.
8. Advise top management immediately of any serious or potentially serious accident for which treatment has been required.

Contents of first aid box

Advice on first aid facilities may be obtained from state health departments. Legislation and regulations in each state should be consulted, as in some cases a list of minimum contents is set out. At the same time, there may be peculiarities of the individual work location which mean that extra items are needed. This information can be obtained from the health and safety survey.

¶5-490 Responsibilities of employees and other parties

Employees also have responsibilities under OHS legislation. Generally, the law requires them to take reasonable care of their own and others' health and safety, as well as to cooperate with systems established and not misuse anything provided to protect health and safety. They should also report hazards or OHS problems to their supervisors, and follow safe working procedures.

In addition, they should understand and contribute to the risk management process, by taking part in the identification and assessment of risks, and by helping to monitor the effectiveness of control measures.

Other parties with OHS roles and responsibilities include designers and manufacturers of plant (that is, machinery, equipment and appliances) to be used at work, suppliers of hazardous substances to be used at work, and people installing or erecting or dismantling workplace plant or equipment. These parties must ensure the plant or substances they design, manufacture, supply, or install are safe to use when properly used. They must also make available information about the safe use of the plant or substances they supply. For example, manufacturers of equipment must provide manuals or instructions setting out information such as maintenance required for safe use, and suppliers of workplace chemicals must make available material safety data sheets which provide specified information about precautions for safe use, first aid and other related matters.

Occupational health and safety professionals such as occupational physicians, occupational health nurses, occupational hygienists, ergonomists and others can also assist with many aspects of OHS risk management. See Appendix C for information on the types of OHS professionals, their roles and the services they provide, as well as when and how to engage them.

OTHER CONSIDERATIONS AT THE PLANNING STAGE

¶5-520 The effect of job design

It is important to understand the potential effects of job content and job design in order to identify any risks arising from these aspects of work.

Employees may quickly lose interest in a boring, mundane, monotonous job (particularly one over which they have little control or discretion). The likely results of such situations are dissatisfaction with a job and alienation. This in turn may lead to consequences such as lack of attention, tiredness, daydreaming/distraction, lack of care, errors and even sabotage. The adverse effects on safety, costs and productivity are obvious, and they can be compounded by other unsafe or unhealthy aspects of the working conditions.

Many factors may contribute to unsatisfactory job design and content, including:

- lack of control over the job, a particular problem with assembly-line situations and where technological enhancement results in "de-skilling" and/or where the operator has to meet the pace set by the new technology
- inability to use potential and initiative
- lack of "relevance" of work — the worker is unable to see the point of the work or its end product
- lack of variety in work tasks, a problem that can often be overcome by introducing job rotation
- a mismatch between job needs and opportunities with individual ability, skills and aspirations
- social isolation, and
- lack of feedback and recognition on the job.

Basically, these add up to a conflict between the needs of the employee and the needs of the organisation and may result simply from an unimaginative approach, or through management being unaware of potential problems. Consultation with employees through health and safety representatives or health and safety workplace committees on matters such as the implementation of changes may obviate many such problems.

When designing jobs and allocating employees to them, attempts should be made to align individual and job needs as much as possible. Popular techniques include job enlargement and enrichment schemes, attitude surveys, forms of employee participation in decision making, quality circles and the use of autonomous or semi-autonomous workgroups. Remember that the more interested and satisfied an employee is with the job, the more chance there is that the job will be performed efficiently and *safely*.

¶5-530 Ergonomics

Ergonomics was defined at ¶1-200. It can be critically important for preventing back injuries and other musculoskeletal disorders to recognise the impact of ergonomics on workers' wellbeing. (See Chapter 10 for more information on managing the risk of back injuries and other strains and sprains, as well as overuse injuries.)

As indicated by the definition, ergonomics covers a wide range of workplace aspects — especially location, comfort, layout and ease of use. In any health and safety survey the layout of each workplace should be considered. The employees concerned can usually provide information on aspects of design or layout which are not user-friendly.

As ergonomics is a very complex science, it is advisable to seek expert advice when purchasing or installing new equipment, instituting new work processes or changing the workplace layout.

Personal observation of and interviews with employees prior to introducing change or purchasing new equipment provides useful information to highlight deficiencies — a health and safety representative and/or committee provides a mechanism for consultation for such purposes.

¶5-540 Catering for employees with disabilities

An important part of health and safety is to ensure that the workplace is safe and convenient for any current physically disabled employees (or suitable for future ones). This refers to a wide range of disabilities and is a consideration that may become more relevant with an increasing emphasis on the rehabilitation of employees. As what will be required varies according to the circumstances of each individual, it is only possible to provide general guidelines on what needs to be considered.

Important information to be aware of includes state building regulations and AS 1428: *Design for access and mobility*.

Items to be considered include:

- building layout — parking, ramp entrances, doors, lifts, washrooms, level entrances and adjacent floor levels
- work layout — desks, shelves, drawers, chair adjustment, work-benches, and
- work equipment — a wide range of equipment specially designed for disabled people (such as those with sight disabilities or limb amputations) is available.

Advice is available from manufacturers of equipment and organisations involved with disabled people. It can also be helpful to consult the employee, who is in a very good position to know what he/she can and cannot do, and what assistance is needed.

Chapter 6

IMPLEMENTATION

HAZARD IDENTIFICATION

¶6-100 Overview

To successfully manage occupational health and safety (OHS) risks, it is important to understand the scope of matters that need to be taken into account when identifying hazards and risks. Analysis of the causes of major industrial accidents over past decades reveals that time and again, some key risks were overlooked, with disastrous consequences. If the spectrum of risks considered is too narrow, or if likely combinations of risks were not foreseen, risks can remain uncontrolled. It cannot be emphasised too strongly that OHS failures, like business failures in general, often result from failures to identify risks which can be all too obvious with hindsight.

For risk identification procedures to be systematic and effective, it is of course vital to allocate sufficient resources in terms of time, personnel, specialised skills and equipment if needed. Areas of responsibility and accountability also need to be defined, documented and communicated, and people undertaking hazard identification need to receive suitable training so that they are competent to carry out the task.

Techniques for identifying hazards

Methods of identifying hazards and risks to health and safety have already been touched on in Chapter 3, "Risk Management". In any organisation, the methods chosen should reflect the nature of the enterprise and the most effective and feasible way of meeting the objective.

Hazard identification processes should take into account the situations or events or combinations of circumstances that could result in accident, injury or ill-health. A combination of methods may be required. The approach chosen should be selected and modified as appropriate, depending on the needs and circumstances of the industry and the nature of the work activities. Some key tools and techniques are listed below.

Observations of the workplace

Management and workers should note unsafe situations or practices and bring them to the attention of supervisors responsible for rectifying the problem.

Analysis of information about past performance

Records of first aid treatment, injuries and work-related illness can be used to indicate hazards. The results of accident investigations can also point to underlying weaknesses in the OHS management system, such as inadequate training, which can increase the risk of incidents or injuries.

Consultation with workers

Workers' on-the-job understanding of how the work is performed often give them valuable insights into the risks of the various tasks. Consulting workers and involving them in risk identification processes will often reveal problems which should be addressed. Minutes and agendas of consultative meetings as well as complaints from employees can also help to highlight health and safety risks faced by workers.

Workplace inspections

Formal inspections of the workplace by managers, supervisors, health and safety representatives and committee members are useful hazard-spotting exercises.

External information

Codes of practice, safety guides, standards and other such material available from OHS authorities, unions, industry bodies and other sources offer reliable pointers to health and safety risks. Other important sources of data include national accident, injury and workers compensation statistics which provide information on the numbers and frequencies of accidents related to type, activity and industry sector. Similarly, information provided by manufacturers and suppliers of equipment or substances for use at workplaces should be used to identify hazards. Operating manuals for machines and material safety data sheets for chemicals generally include important safety information which indicates foreseeable risks.

Monitoring of workers' health and/or the working environment

Measurement and testing by OHS professionals may be necessary with some types of hazards to establish whether they present a risk to workers' health or safety. For example, noise levels may be measured using special equipment, or samples of gases taken and analysed to test for airborne contaminants. Biological monitoring may involve testing samples of workers' blood or urine to test for levels of substances such as lead or pesticides.

Other techniques include the use of checklists, brainstorming and analysis of possible scenarios. If all risks are to be uncovered, it is important that the approach to hazard identification is both systematic and comprehensive.

Hazard identification at the planning or design stage

It is essential that all proposed changes — such as changes to equipment, systems, substances used, work practices, legislation and standards — be evaluated for OHS implications. Many problems can be prevented by considering any health and safety ramifications *before* changes are made.

Documentation

Documenting the hazard identification process provides a basis for ongoing monitoring of risks and creates a record of the organisation's diligence in identifying foreseeable risks of injury.

Survey

One important technique is to conduct a thorough survey by trained personnel to determine existing practices, standards and conditions in each work activity. Employees can also play a part in this survey. The survey can be taken either of the whole organisation at once, or of one section at a time.

The scope of this initial survey has a "one-off" character as distinct from the ongoing activity of health and safety auditing and more routinely undertaken monitoring. (Ongoing auditing of health and safety is discussed in Chapter 7.) Information gathered in this survey can form the basis of an organisation's ongoing control of health and safety.

An initial survey should include consideration and/or measurements of the following:

- lost-time and minor injuries (obtained from workers compensation claims, injury report forms and first aid records) also showing cost differences, as the most frequent incidents need not be the most costly, and vice versa
- sick leave days lost (from human resources/pay office records)
- damage to plant and equipment (from maintenance records)
- workers compensation and insurance arrangements and costs (from the accounts section)
- the background of employees in terms of such aspects as personal disabilities (such as the need for prescription spectacles or the incidence of respiratory ailments), cultural background (whether English-speaking and whether any other languages are spoken) and extent of work experience. In some situations, age distribution may also be relevant (several studies have suggested that the inexperienced, usually younger, worker is most at risk of injury, particularly during the first six months on the job)
- job design (from observation and discussion consider safety, ergonomics, training and work activity layout)
- compliance with legislative requirements
- types of equipment and raw materials used by the organisation, with emphasis on toxic and flammability levels and other possible health hazards (consider the quality of information available on hazardous substances — material safety data sheets are discussed at ¶10-240)
- storage and disposal of waste, including any stockpiling of dangerous/ hazardous materials
- preparedness and provisions for fire safety and other types of emergencies (emergency control plans, alarm systems, fire extinguishers, emergency evacuation drills, trained employees, identification of areas where fires are likely to start), and
- current risk control measures, including a list of protective clothing and equipment used at the workplace.

All of these sources and records should provide relevant information about current organisational occupational health and safety practices. From this it should be possible to compile an overview of the accident and injury trends and probable areas of risk that require closer attention, supervision and control.

¶6-120 Some basic questions to ask

To form a starting point for identifying risks of accident and injury, some basic questions are suggested below. Note that each organisation should formulate a range of specific questions in relation to the type of business it is engaged in.

- What kinds of accidents/injuries/damages are occurring, or typically occur in this type of work?
- How frequent are the various types of accidents/injuries/damages? What are the costs incurred?
- Where are the accidents/injuries/damages occurring?
- Which occupations of employees are related to the accidents/injuries that are occurring?
- Which accidents/injuries/damages cause the greatest time loss?
- What objects and/or substances are most associated with accidents/injuries/damages?
- What factors are associated with accidents/injuries/damages, and is there a recurring item?
- Are there documented safe work method statements, or safe operating procedures, and are these used in practice?
- Is there a system of accountability and are clear guidelines in place in all relevant sections?
- What are the procedures used for accident reporting and investigation? How effective are these?

The information that is collected may reveal weaknesses within the management control system that require immediate attention. So, it may provide a factual foundation on which an OHS risk management program can be developed to meet specific needs. As the collection of this information may be a considerable task, some thought should be given to a system of storing the information so that it may be easily referred to and updated in future.

Note that the above questions relate only to the risk of accidents. As well as employee safety, it would be wise to consider all other areas in which the organisation has a duty of care, such as employee health, product safety, environmental safety and the safety of the general public.

Who should undertake the health and safety survey?

If the survey is done in-house, it should be carried out by people with training in safety and/or health matters. Even if the organisation has staff with full-time involvement in these functions, an assessment should be made to establish whether they have the skills to carry out this type of activity. If they do not, training or retraining should be arranged first.

Some companies use outside consultants for some surveys and their own employees for others. This depends on the nature of the survey and the availability of on-site expertise. The combined approach has the benefit of achieving a balance between specialist or expert input and the development of a company's own employees.

Health and safety professionals are considered in Appendix C.

Organisations can arrange for different types of surveys from outside sources:

- insurance brokers or insurance companies with whom the organisation has a policy
- private health and safety consultants, such as specialists in noise control
- various federal and state government departments (see Appendix A), and
- some safety equipment suppliers.

Role of employees in the survey

Where health and safety representatives and/or committees are established at the workplace, it will be useful to involve them in the survey. This may take the form of consultation in the design of the survey, or involvement in carrying out the survey. As employees, representatives/committee members will be able to provide information on problem areas and common occurrences. Further, involvement of representatives/committee members should lead to a greater level of interest on the part of the other employees.

Each state and territory makes specific provision for a consultative style of workplace inspection in its occupational health and safety legislation. The relevant provisions in each state/territory are summarised at ¶2-140. In general, health and safety representatives or committee members are able to inspect the workplace on giving reasonable notice to the employer. In some states provision is also made for a representative to accompany an inspector on a tour of the workplace.

Outside any specific legislative provisions, it is suggested that members of a health and safety committee will need to undertake an inspection of the workplace to familiarise themselves with details of its layout, procedures, problems, hazards and other relevant matters.

Identifying specific organisational problems

While it is both possible and useful to conduct a health and safety survey along a broad range of general guidelines, many organisations will have their own specific health and safety problems that may require further study and/or outside professional assistance.

Some examples include the following:

- particular types of work and work processes (such as abrasive blasting, spray painting, mining and tunnelling)
- working with dangerous or harmful substances (such as toxic or flammable chemicals), radiation and hazardous dusts (such as asbestos or silica)
- unusual working conditions (such as heat or cold exposure) or animal communicated diseases (such as brucellosis), and
- language or other communication problems (such as mixed ethnic groups, noisy situations or partially deaf workers).

In other cases the initial health and safety survey may alert an organisation to particular problems at the workplace. Either way, the particular problems will require special attention and should be considered in the planning process. A number of the most common health and safety problem areas are considered in Chapters 9, 10 and 11.

¶6-150 The health and safety audit

In financial management, there is a projected budget for which capital funding has been allocated. Receipts for expenditure are measured against that known amount to determine whether the organisation is on target for the particular period. A health and safety audit is a similar management tool to measure performance and costs. It is the next stage of development from the checklist system. The latter tells "what to look for", but the audit requires that an item must be converted to *a measurable standard* with its own rating procedure.

A health and safety audit is a systematic and periodic review of the whole OHS management system, including the policy and programs used to promote OHS and prevent workplace accidents/incidents and work-related illness. An OHS audit measures the performance of the whole system to determine if the OHS systems are in place and if they are working, that is, if the control strategies are effective.

Health and safety auditing is an ongoing activity. The initial health and safety survey can be used to establish standards against which future audits can be measured. Performance measurement is discussed in Chapter 7.

RISK ASSESSMENT

¶6-200 The process

As outlined in Chapter 3, risk assessment is the overall process of estimating the magnitude of the risk and deciding what action to take. The process involves:

- estimating the probability or **likelihood** of an accident occurring
- estimating or calculating the **severity** of the potential consequences, and
- based on these two factors, assigning **priority** to the risks for risk control.

Estimating likelihood

The method of determining the likelihood or probability that injury, illness, damage or loss will occur as a result of hazards will vary according to the type of workplace and operations involved. Determining likelihood is often more difficult than deciding on the severity of the potential harm. Many assessors consistently underestimate the likelihood of hazards causing harm. This is often because their experience of real accidents is somewhat limited. A basic system of evaluating likelihood is outlined below.

Review information

In considering likelihood, it is important to review any information which was gathered during the hazard identification stage. This may include, for example:

- hazard identification checklists (which will indicate the factors taken into consideration)
- hazard identification record forms (which will provide valuable information on circumstances surrounding the hazard together with comments of the identification team or individual)
- the organisation's accident, incident and first aid records (which should reveal trends or frequencies of injury)
- accident/incident investigation reports
- the organisation's workers compensation records
- plant maintenance and breakdown records (such as service books)
- work systems and procedures documentation
- OHS policies — both general and specific
- employee training records, and
- operators' manuals, material safety data sheets and equipment instruction booklets.

Rating the likelihood

The following scale may be used as a model for rating likelihood using a basic numerical range of 1 to 10:

- Certain or imminent (10)
- Highly likely (8)
- Likely (6)
- May happen (4)
- Unlikely (2)
- Highly unlikely (1)

Factors influencing likelihood

When deciding which of the above values to attribute to a particular risk, the following factors should be taken into account:

- the number of times the situation occurs (for example, how many times a day a construction worker is exposed to the risk of back injury through manually handling heavy bags of cement)
- the number of people exposed to the hazard (for example, the number of nurses caring for patients who are potentially aggressive)
- any special characteristics of the people involved (for example, a person who has had a prior back injury may be more likely to sustain another; a person who is already sensitised to solvents may be more likely to sustain health effects than another worker)

- the position of the hazard (issues such as distance from a source of loud noise may influence the likelihood of noise-induced hearing loss)
- distractions (such as time pressures or workplace conditions which may influence careful undertaking of a task — for example, hectic activity in a hospital emergency area when a nurse is taking a blood sample could impact on the likelihood of a needlestick injury)
- duration of exposure (for example, how long the person is exposed to a source of very loud noise)
- quantities of materials or multiple exposure points involved
- environmental conditions (conditions which may increase the likelihood of an incident occurring — such as water in the vicinity of an electrical hazard)
- competence of people involved (for example, adequate training and experience of workers), and
- condition of equipment (defective equipment is more likely to be involved in an accident — for example, a cracked guard on the power take-off shaft of a tractor may actually increase the likelihood of accidents rather than preventing them).

Consider current controls

A further important factor to take into account is the effectiveness of any control measures already provided. It is also necessary to consider the possibility of current control measures not being used due to issues such as:

- lack of training or supervision
- failure to replace controls following cleaning, maintenance or repair work
- difficulty or awkwardness in using or working with controls, and
- complexity of controls.

The efficiency of the control measures in preventing harm will affect likelihood. If control measures are found to be a factor, a check of the organisation's policies and procedures may be required to discover the original purpose and intended use of the controls.

Realities of work practices

It is essential to consider how work is **actually** carried out in the workplace and to base the assessment on this, rather than assessing likelihood based on how the work is **supposed** to be carried out. The two may vary significantly.

The assessor or assessment team carrying out the risk assessment (and this should include key personnel, such as the workers involved and their supervisors) may decide at this point that they do not have sufficient information or specialist knowledge to determine whether a hazard is likely to cause harm. In this case, it may be helpful to actually go and observe the work being done, and to undertake appropriate measurements or testing. Otherwise, it may be necessary to seek outside professional assistance.

Decisions about likelihood

The following information is designed as a "rule of thumb" to assist in deciding which value to place on the likelihood of harm occurring, using the values outlined earlier.

Certain or imminent (10)

Harm from a hazard is **certain or imminent** if no control measures are provided at all and if one (or more) of the following factors also applies:

- the hazard is a cause of large numbers of injuries or ill-health in national statistics
- people are exposed to the hazard continuously
- the hazard is difficult to see
- the hazard is encountered in adverse environmental conditions, and
- safety is not considered a high priority by those involved.

Highly likely (8)

Harm is **highly likely** if one or more of the following is true:

- any control measures provided depend on an individual using them on every occasion (for example, personal protective equipment and safe working procedures)
- training and supervision are minimal, and
- if one or more of the factors outlined for "certain or imminent" harm is present.

Likely (6)

The likelihood ranking will be reduced to **likely** if one or more of the following is true:

- any control measures provided depend on an individual using or adjusting them on every occasion
- training is provided and work is supervised, and
- any of the factors given for "certain or imminent" harm also apply.

May happen (4)

Harm **may happen** if one or more of the following is true:

- control measures include measures that do not depend on the operator but can break down or be removed or defeated, and
- there is no defined system of maintenance or supervision of the control measures.

The choice of this ranking may also be appropriate if one or more of the following is true:

- the hazard is a significant cause of injury or ill-health, and
- a number of people are often exposed to the hazard.

Unlikely (2)

The circumstances under which harm may happen (above) will reduce to harm being **unlikely** if one or more of the following is true:

- there is a defined system of supervision and maintenance, and
- adequate training is provided and repeated regularly.

The choice of this ranking may also be appropriate if one or more of the following is true:

- the hazard causes few injuries or cases of ill-health, and
- a few people are only occasionally exposed.

Highly unlikely (1)

Harm is **highly unlikely** if one or more of the following is true:

- control measures are unlikely to break down or to be removed or defeated easily
- there is a maintenance procedure in place, and
- adequate training and supervision are provided.

The choice of this ranking may also be appropriate if one or more of the following is true:

- the hazard rarely causes injury or ill-health
- only one or two people are exposed to the hazard infrequently, and
- there is a very positive safety culture in the organisation.

Estimating severity

The method chosen to determine the severity of the potential injury, illness, damage or loss associated with the hazard will depend on the unique needs and circumstances of the workplace. A basic system is outlined below.

Rating the severity

To determine the severity rating, the assessor or assessment team must make a judgment on the possible outcome of the hazard. The assessor(s) should use common sense as a guide, but further guidance can be provided by information such as accident statistics, manufacturers' data and information gathered during the hazard identification process.

When considering how severe the harm from a hazard could be, it is important to be **realistic**. Almost every hazard could result in death. For example, it is remotely possible that someone tripping over a cable in an office may be killed, but the most probable result is bruising or at worst a fractured bone. If, however, the cable is trailing across the top of very busy stairs a more severe injury, even death, is possible. Similarly, a splinter of wood in the hand could result in blood poisoning and death but minor injury is a more realistic outcome.

The following scale may be used as a model for rating severity using appropriate terminology and a basic numerical range of 1 to 10:

- Multiple deaths (10)
- Single death (8)
- Major injury (6)
- Lost-time injury (4)
- Minor injury (2)
- Delay only (1)

Factors influencing severity

Factors affecting severity include:

- the number of people who may be affected in one incident
- special characteristics of individuals which place them at increased risk (such as lack of experience or medical conditions)
- concentrations of substances
- volumes of materials
- speeds of projectiles and moving parts
- heights and distances
- weights, and
- forces and energy values.

Current controls

When analysing the severity of potential harm, any control measures already provided such as guards, protective clothing, or permits to work should **not** be taken into account — unless they reduce the risk at source.

For example, the provision of safety footwear to the operator of a hover lawn mower **does not** make the blade less dangerous. The blade could still cause major injuries to the hand or ankle and someone who decided not to wear the safety footwear could lose toes. The fitting of a plastic blade, on the other hand, **does** reduce the risk of injury to a minor cut or bruise as the blade is intended to snap off before it can do further damage.

Decisions about severity

The following information is designed to assist in deciding which value to place on the likely severity of harm which may occur.

Multiple deaths (10)

This rating may be applied in situations which have the potential to kill a number of people in one incident, such as:

- work on scaffolds
- work involving confined spaces
- demolition operations
- fire or explosion hazards, and
- use of substances giving off toxic vapours.

Single death (8)

This rating may be applied where the hazard may result in a single death, such as:

- work at heights (over two metres)
- large power press operations
- use of 240 V electrical tools, and
- operations involving forklift trucks, tractors, trucks or other heavy vehicles.

Major injury (6)

This rating may be applied where the hazard may result in major injury, such as:

- work at heights up to two metres
- use of equipment such as circular saws or grinders
- exposure to people with a propensity towards violence, such as some psychiatric patients
- welding work
- dusty operations
- noisy work (where noise levels exceed limits specified in legislation)
- needlestick injuries involving infected blood
- significant manual handling of heavy loads, including people
- work involving excessive vibration, and
- repetitive actions.

Lost-time injury (4)

This rating may apply where there is a potential for injury resulting in time off work, such as from:

- lesser manual handling hazards
- tripping or slipping hazards, and
- corrosive or harmful substances.

Minor injury (2)

This rating may apply to hazards which could result in minor injury, such as:

- handling sharp objects, and
- using mechanical hand tools.

Delay only (1)

This rating involves a negligible OHS risk but may be relevant from a loss control perspective. Quality control issues may be identified that could affect production but will not cause personal harm, such as incorrect specification of materials causing a machine to jam.

For every accident that results in serious harm, there will be many "near misses" or incidents. The difference between the **occurrence** of a near miss and a fatal accident can be largely a matter of chance or luck — whereas the **potential** for harm may be the same in both cases. Sometimes a split second or a few millimetres makes all the difference. Organisations which stringently record near misses and incidents as well as accidents will be in a far better position to determine severity (and likelihood) than those which don't.

Prioritising the risk

Once the likelihood and severity of the potential harm have been rated, it is possible to prioritise the risks based on these two criteria. Prioritising risk is the final step in the risk assessment process.

Any control measures currently applied were taken into account in earlier calculations. Assigning priority involves ranking risks as:

- trivial ("Low" priority)
- adequately controlled ("Medium" priority), and
- not adequately controlled ("High" priority).

A basic system which can be used for making these decisions is to draw a matrix or table such as the sample below, showing likelihood (probability) and severity, and to position each risk within it.

The likelihood/severity matrix						
Severity → X Likelihood ↓	Multiple death (10)	Single death (8)	Major injury (6)	Lost-time injury (4)	Minor injury (2)	Delay only (1)
Certain or imminent (10)	100	80	60	40	20	10
Highly likely (8)	80	64	48	32	16	8
Likely (6)	60	48	36	24	12	6
May happen (4)	40	32	24	16	8	4
Unlikely (2)	20	16	12	8	4	2
Highly unlikely (1)	10	8	6	4	2	1

This enables both the key factors to be taken into account when ranking risks in order of their priority for risk control.

It may happen that a decision cannot be reached using a general risk assessment as the hazard is one for which special expertise or further information is required. In this case, that finding should be recorded, and it may be necessary to undertake some form of testing (such as workplace monitoring or health surveillance of employees) to make a precise determination of the risk. It may also be necessary to use a more specialised risk assessment method.

Specialist risk assessment techniques

In complex hazard situations, such as chemical plants and large-scale manufacturing operations, the application of the basic risk assessment techniques outlined above may not be sufficient to take account of the diverse permutations of risk factors present. There are many specialist risk assessment techniques which can be used to better understand the implications of major hazards and complex environments.

Techniques commonly used include:

- Hazard and Operability Studies (HAZOP)
- Failure Modes and Effects Analysis (FMEA)
- Fault Tree Analysis (FTA)
- Event Tree Analysis (ETA), and
- Human Reliability Assessment.

To competently undertake any of these specialist techniques, specialist training and experience is needed. Further information about these techniques may be found in CCH's publication *The Hands On Guide — Risk Management*.

RISK CONTROL

¶6-300 Selecting options for risk control

As mentioned in ¶3-140, implementing strategies to control risks begins with the identification of the various control options. Risks can be avoided where this is feasible, or changes can be made so that adverse consequences are less likely or can be minimised.

Guidance and advice on control options are provided for a number of common risks in Chapters 9, 10 and 11 of this book, and further information is available from other CCH products and OHS authorities' websites (see Appendix C). Information from sources such as these should be collected and evaluated, and the most suitable measures chosen. Ideally, key personnel including relevant workers and their supervisors should participate in this process. In some cases it may be advisable to seek advice from OHS specialists (this is covered in more detail in Appendix C).

Selecting the most appropriate way of addressing a given risk involves balancing the probable effectiveness against practicability and cost, though for health and safety risks, legal and social obligations override simple financial cost benefit analysis.

The hierarchy of risk control

Risks should be controlled using the hierarchy of risk control, outlined in Chapter 3. The preferred order of control measures is based on their effectiveness. This means that elimination should be the first control method to be considered. If it is not possible to eliminate the risk, methods of minimising the risk should be considered, based on reasonable practicability.

In many cases, a combination of control measures should be devised, as appropriate to the nature of the risk. Control measures should be documented for further follow-up. Administrative control methods such as training, the establishment of basic workplace health and safety rules, and the development of safe operating procedures are often used in combination with other measures.

Risk control methods are not developed in isolation from the work methods used in a workplace. They must become an integral part of the operation of all work systems. To ensure that control measures and work methods are entirely compatible, it is recommended that the relevant control measures are written into all standard operating procedures.

¶6-320 Standard operating procedures

Standard operating procedures (SOPs) (sometimes referred to as work method statements) are step-by-step guides to carrying out a task properly. These will discourage the taking of risky short cuts. They should be written in plain English; workers should be trained in their application and supervision should be adequate to ensure they are adhered to. Once all identified hazards and controls have been included, the procedures become a blueprint for safety as well as efficiency by standardising for all employees the way in which a job is undertaken. Standard operating procedures should be developed by a team, including:

- experienced employees who carry out the task
- the line manager/supervisor for the work area
- a management representative
- the risk management program coordinator, and
- the OHS manager (if separate from above).

In producing new SOPs, job method statements are reviewed, jobs may be redesigned and new hazards may be uncovered. By including employees in the process of developing SOPs, they become participants in the process of identifying hazards, assessing risks and finding better, safer and more cost-effective ways to do the job.

Standard operating procedures are also an ideal on-the-job training tool. In addition, they provide accident investigators with a detailed guide to the work being undertaken prior to the accident.

Standard operating procedures should be reviewed when:

- an accident occurs

- a work method is changed
- a job process is changed, and
- risk control measures change.

¶6-330 Health and safety rules

Another common outcome of the risk control process is a set of basic OHS rules which must be followed by all employees.

All employees will need to be informed of the rules by both receiving them and by prominent displays in the workplace.

The rules should be basic, supplemented by more detailed operating instructions when the latter are needed.

Where possible, observance of the rules should be contained in an induction document as a condition of employment, with failure to observe them rendering an employee liable for further action, such as retraining or, where appropriate, discipline.

Scope of rules

Rules should cover aspects such as the following:

- "housekeeping"
- use of machinery
- use of protective equipment
- maintenance procedures
- accident reporting
- first aid attention
- fire prevention
- electrical restrictions
- use of tools and equipment
- authorised entry and usage (certificates, licences)
- smoking, alcohol and drugs
- prohibition of horseplay and misuse of equipment, and
- any other rules applicable to the organisation's work.

Example

The following example shows a typical set of basic health and safety rules. Any set of rules will need to meet the needs of the organisation or work area.

Basic health and safety rules

These basic health and safety rules apply equally to every person working within this organisation.

Any breach of one of these rules may result in serious injury to one or several people. Therefore, each supervisor has the duty and authority to take appropriate disciplinary action after any deliberate violation of any of these rules.

1. **Know and observe departmental health and safety rules**

Each department or section has its own specific operating rules and procedures in addition to these basic health and safety rules. Each employee must be aware of, and follow at all times, these rules and procedures when performing the job.

2. **Observe all warning signs and danger tags. Keep all safety guards in place**

A warning sign, danger tag or safety guard is put there to ensure your protection from recognised hazards. This rule is necessary to protect everyone.

3. **Use protective equipment where required**

Employees must make proper use and take proper care of the personal protective equipment provided for on-the-job use. Any faults in the condition of any of the equipment should be reported to supervisors.

4. **Report all injuries or potential hazards immediately**

A first aid station is provided for treatment of injuries in all work sections. Prompt treatment for even small injuries is the best way to prevent painful and costly complications. All injuries must be reported as soon as possible to the first aid station and details of accidents reported in the accident book. The supervisor must also be notified.

5. **No smoking on the job**

Because of several highly flammable chemicals used in work processes, flames and sparks must be carefully controlled to avoid fire or explosion. Smoking is permissible only in designated areas.

6. **No alcohol and drugs on the job**

People under the influence of alcohol or inappropriate drugs can make errors in the performance of their work, as well as presenting a risk of accidents or injuries to themselves or others. Your work requires the maximum degree of alertness and good judgment. If your doctor prescribes drugs to treat a medical condition, you will need to advise your doctor of your work environment so that he or she can determine whether it is safe to continue working whilst taking that medication.

7. **No horseplay**

Serious injuries and incidents can result from horseplay or practical jokes.

Signed..................... Date.....................

(Chief Executive Officer)

¶6-340 Recruitment and selection of employees

Another administrative aspect of risk control consists of ensuring that the process of recruitment and selection of new employees, as well as the formulation of job descriptions and performance criteria, take account of the need to manage health and safety risks. The job specification for each position should contain the particular skills, knowledge and abilities that are required by a person to be able to successfully undertake the tasks involved in the job in a safe manner. There may also be particular statute or award provisions applying to junior employees, for example.

Examples of requirements could include physical fitness, eyesight, mental alertness, height and weight (only where relevant), education, language, possession of particular skills and qualifications, and general health (see medical examinations at ¶6-350). These requirements should only be included where they can reasonably be considered to be essential for a position. Where this is not the case, including such a requirement in the criteria for a position could amount to a breach of equal opportunity legislation.

Job descriptions should include more general safety statements. For example, a job description could include the following as a required duty:

> "Comply with safe working procedures. Monitor and maintain a safe working environment and report any health or safety hazards to area supervisor."

The job description for a storeman could include the following:

> "The worker will be expected to utilise safe manual handling procedures in accordance with safe operating procedures and training."

Statements such as these, while being general, will emphasise the importance the organisation places on health and safety. Similar statements can be placed in advertisements for positions.

Performance criteria for supervisors and line managers should also incorporate measurable performance indicators relating to health and safety objectives and targets. These will provide management with feedback on what is happening. They can be outcome-based, reflecting system or operational performance (eg rate of injury), or they can be input-based (eg number of inspections conducted, instruction and training provided or audits performed). See Chapter 7 for more discussion of performance measurement.

¶6-350 Medical examinations

Reasons for medical examinations

An organisation may undertake medical examinations of its employees for a variety of reasons.

1. Pre-placement

A pre-placement test may be appropriate to determine an applicant's suitability for a particular job. For instance, has the person sufficient physical strength if the job requires heavy work; is eyesight sufficiently keen (or appropriately corrected) to meet the requirements of assembly of small parts; is the person unaffected by colour-blindness if working in some industries such as electrical trades; is the person prone to allergies or bronchial trouble if working in particularly dry, moist or dusty atmospheres? The particular requirements will obviously vary from industry to industry, and are designed to protect both the employer and the employee.

The examination, if part of the selection procedure, should be held near the end of the selection process, so that only applicants who appear otherwise suitable are tested. This has the advantages of saving time and money, and placing the test as near to the person's entry into your organisation as possible. The examination is a record of a person's fitness at the time of hiring and further examinations will show deviations that could be of assistance in workers compensation claims or common law suits. Pre-placement examinations should also be used where employees are transferred within an organisation.

Note that the aim of a pre-placement examination should not be to exclude persons who fail to measure up to a rigid health standard but to check their suitability for the type of work involved. It may be that people with certain medical disabilities would still be suitable for other jobs within the organisation.

In recent years there has been some controversy over pre-placement or other testing to screen for alcohol or drug use. This is becoming more common in North America, but is not well-accepted in Australia. While there may be a case for screening in certain jobs (airline pilot, train and coach drivers), cost and privacy considerations make it less acceptable in other situations. There are also problems with testing because it can be unreliable as an indicator of a person's ability to function safely. Tests of function such as simulation tests can be a much more useful and valid way of assessing a worker's ability to perform tasks safely. See ¶9-670 for further discussion of testing for alcohol or other drugs. The provision of employee assistance programs (see ¶9-650) may be a better alternative than testing, as a means of managing drug and alcohol issues.

2. Periodical examinations of employees

Periodical examinations may be appropriate and are conducted most commonly on an annual basis. These examinations can be either of all employees or of subgroups (such as executives). The purpose of these examinations can include both general health (blood tests, electrocardiograms, radiography) or tests for specific occupational hazards (such as noise-induced hearing loss or exposure to potentially harmful substances. In most cases these examinations will not be necessary.

3. Entry to superannuation and disability schemes

Superannuation funds and disability schemes may require a medical examination before accepting a new member.

4. Workers compensation

Medical examinations may be required in workers compensation cases to determine fitness to resume usual work activity, or for placement into alternative work situations providing meaningful work tasks for people whose injuries impose some restrictions on the type of work they can manage.

5. Testing of employees for HIV

In practice, there are very few industries where there is any possibility of transmission of the AIDS virus at the workplace. With this in mind there seems to be little justification for testing for HIV infection at the workplace. In any event, testing should not be performed without an employee's consent, and any test results should be treated in the strictest confidence. An employer who dismissed or transferred an employee because he/she was HIV positive may be guilty of unlawful discrimination and be liable for damages.

> Medical examinations should not be perceived by management or employees as a means of penalising employees for any health problems they may have developed, but as an effort to ensure the continued good health and safety of all employees.

A major advantage of regular examinations is that any problems, particularly ones partially or wholly due to factors at the workplace, may be identified and corrected before they become serious. Statistical data from examinations could also be used in some cases as a means of approaching and reducing the causes of health problems.

If the organisation lacks the facilities to conduct or arrange medical examinations of employees, it may still be a good idea to encourage employees to voluntarily arrange their own medical examinations at regular intervals.

Confidentiality issues

The collection and use of medical information on employees is a very sensitive issue.

Employers should ensure that medical reports are kept confidential, and that medical records are not kept any longer than a legitimate use for them exists. For example, the NSW Privacy Committee recommends that employers should destroy medical reports of every unsuccessful applicant upon request of the applicant or within two years after application for employment is made, unless exceptional circumstances exist or the applicant requests that the report be kept for a longer period. Generally, organisations are required to adhere to the National Privacy Principles, which set out minimum standards on how organisations should handle personal information. More information about privacy requirements is available from the Office of the Federal Privacy Commissioner.

The employer is not the appropriate person to discuss the contents of a medical report with an applicant. If details of medical reports are requested by an applicant, the person should be advised to contact the examining doctor or qualified staff. If the employer's consent is required for the release of details, it should not be denied.

Note that a neutral stance by the doctor will be essential. The doctor must never be perceived as catering for the interests of management. A rigid code of ethics for communication of information will need to be worked out and adhered to.

If medical records are to be retained by employers, access to them must be strictly controlled. It would seem better, however, for any detailed records to be kept by the doctor and general advice given by the latter to employers.

MONITORING AND EVALUATION

¶6-380 Monitoring and evaluating risk controls

Once risk controls have been put in place and any relevant information and training provided, it is important to check that the controls are achieving their desired purpose. It is not uncommon for controls to be unsatisfactory in some way, or to inadvertently introduce a new hazard. Alternatively, circumstances can change and the control can be rendered ineffective by some other factor that was not foreseen, or by changes in legislation or standards. Therefore controls should be regularly checked for their effectiveness and adequacy, and these checks should be documented. See Chapter 7 for further information about evaluating risk controls.

TRAINING AND INFORMATION

¶6-400 Why train?

A key requirement of OHS law is that management must provide adequate training. A disproportionately high number of serious injuries are suffered by either new employees or those commencing a new or different work activity, or on returning to work after leave.

The basic aim of health and safety training is to impress the principles of health promotion, risk management and safe behaviour on employees in such a way that they will apply these principles to their work. The training required will vary according to position held and type of work performed.

A structured health and safety training system should reduce the number of injuries at a workplace, and should be evaluated against that criterion. Associated benefits will be an increase in productivity and a reduction in the costs associated with injury and disease. Further, a formalised training system emphasises an organisation's commitment to a safe workplace; it creates a "culture of safety" and helps the organisation to comply with legislation. It also facilitates the assessment of an employee's competence as required under common and statute law.

¶6-420 Induction training

Induction training is usually the first introduction to an organisation. It is a combination of a formal introductory session and basic on-the-job training that can be conducted by a supervisor. Parts of a program should also be used when an employee is transferred into a section/department from another within the organisation, as they are "new" to that area.

Induction training should include:

- general instruction on the workplace and work performed
- notification of the organisation's health and safety policy and any health and safety programs in place
- information on risk management and expectations of the employee in terms of participation in the process
- advice of health and safety rules
- issue of, and advice on use of, protective clothing and equipment
- advice on particular hazards at the workplace
- response in the event of fire or other emergency
- instruction in necessary skills, such as manual handling, machine operation
- advice on first aid facilities and personnel
- procedure for workers compensation claims, and
- advice on, and identification of, the health and safety representative and/or safety committee.

¶6-430 Continuing training

Effective training is based on needs rather than mere routine. It may be appropriate to target a group of employees who work in a particularly dangerous area, or where statistics reveal a problem, or where a new process is introduced.

The content of any training should be documented and periodically reviewed. Records should be kept of who has received training, and of the content of that training.

Note that there are specific outside courses available on particular subjects such as manual handling, fire safety and hearing conservation. Training organisations can be contracted to provide such courses, either on site or away from the workplace. Another option is to contract a training organisation to design an on-site course specifically to suit the needs of a particular organisation. In some areas, such as crane driving and forklift driving, legislation requires operators to hold an appropriate certificate.

Preparing training

The first step in preparing any training program is to undertake a training needs analysis to determine who needs to be trained, and what the content of the training should be. To assess the needs for OHS training, it is necessary to analyse work tasks in terms of the knowledge and skills required to carry out the tasks safely and competently. Other factors which can be considered in order to identify training needs include:

- records of work injuries and illness
- damage to plant and equipment
- absenteeism
- records of hazard identification and risk assessment processes
- legislative requirements, for example, certification for operators of particular types of industrial equipment such as forklift trucks.

To be effective, training needs to be oriented to the needs of the individuals to be trained, as well as the needs of the organisation. This is not only a matter of the content of training, but of the trainees' needs in terms of language, literacy, numeracy, levels of existing knowledge that can be assumed, and the level of detail needed.

If training is proposed to address an OHS problem such as large numbers of back injuries, the first consideration is whether the problem can be resolved by training alone, or whether other risk control measures are also called for. If training can assist, it is then necessary to define the outcomes/objectives required. These outcomes must be realistic, and at the end of the training program it must be possible to determine whether they have been met. The desired outcomes should be communicated to the trainees at the commencement of training.

Once the outcomes have been set, a training program that will achieve those objectives can be designed. Classroom training or on-the-job training are two alternatives. Other approaches include a supervised project undertaken at work, observation of other employees performing the task (preferably followed by a "try out" of the task by the trainee), or a self-directed group exercise.

Different levels of employees

Different training needs for different levels of employees are set out below.

Employees

All employees need to be sufficiently informed about safe methods of performing the work in which they are engaged. They should understand their role within the organisation's overall health and safety policy, and in the risk management process.

The actual level of training required will depend on the danger and complexity of the tasks performed. An indication of the need for training can be obtained from records of the hazard identification and risk assessment processes, as well as the safety performance of each group of employees.

Non-English-speaking employees

The fact that communication is an acquired skill, even when instructor and trainee normally speak the same language, underlines the necessity for special measures when they do not. Where there are employees who do not speak or understand English well, safety problems can arise if they do not understand instructions and are afraid to admit it. Some words can have similar sounds but totally different meanings.

Measures to remedy this situation include appointment of bilingual employees as interpreters, and written instructions in other languages. Both methods are appropriate, although both have potential problems. Interpreters may tend to develop into unofficial supervisors although they are not suitable in other ways. Written instructions fail when ethnic employees are illiterate or only partially literate in their own language. The overriding concern is to ensure that all employees understand the safety instructions that are given. Demonstrations and the use of pictures or videos can assist in imparting the safety message where there are language problems.

Supervisors should make allowances for the difficulties (and hence the added risks) encountered by workers from a non-English-speaking background. Above all they should ensure that instructions are given in the simplest terms and in the clearest way possible.

Classes for workers to learn English can be arranged at work locations. They are usually arranged partly in work time and partly in employees' time, with the human resources department coordinating the classes. In New South Wales, Victoria, South Australia and Western Australia, the Adult Migrant English Service provides English language programs and related services for people from language backgrounds other than English. Note that some awards, most notably in Western Australia, contain a provision requiring an employer to allow migrant employees to attend English language classes during work time without loss of pay, and in Victoria a code of practice requires the dissemination of information in appropriate languages.

Health and safety representatives/committee members

Training is crucial to enable health and safety representatives or committee members to carry out their roles effectively. This is discussed at ¶5-475. Note that the legislation generally provides that an employer must grant a representative or committee member time off without loss of pay in order to attend accredited training courses.

Supervisors

The supervisor's control over the workplace, its people and resources necessitates an understanding of risk management and accident causation. Health and safety must be accepted as part of the supervisor's job in the same way as all other duties and activities.

For these reasons, there is likely to be more emphasis on formal health and safety training courses, run by either the organisation's training section or an outside organisation such as a government department (see Appendices A and B). It is also important to remember that supervisors will be responsible for a large part of the health and safety training of their subordinates. Therefore, they need skills that include training techniques, as well as health and safety management, report writing, investigation methods and motivational techniques. This may involve their attendance at courses, or their involvement in reinforcing the messages delivered to their staff via training they have received.

Supervisors may also find it beneficial to belong to some outside health and safety organisation to keep abreast of latest developments. Finally, they may benefit from private research and should be on the circulation list for health and safety journals and related publications received in the organisation.

Senior managers

The approach for senior managers will be similar to that for supervisors, particularly in smaller organisations. Health and safety rules and policies should apply to everyone from the highest level of management down. Therefore, health and safety rules and procedures should be included in the induction and training of senior staff.

Managers are likely to have less direct contact with on-the-job matters than supervisors, so their training will focus to a greater extent upon developing strategies to protect health and safety, and on reports/statistics and monitoring. In many cases, it will help to provide information and results from a viewpoint that emphasises costs and profits. Senior managers should make regular visits through their entire area of control and show interest in the health and safety function by discussing it with supervisors.

Information

Training and information can be seen as part of a continuum which includes training, information, instruction and supervision. While training involves the transfer of knowledge and the systematic development of a person's skills, instruction involves directing people in the conduct of specific tasks. Supervision involves taking responsibility for, and directing, the activities of others.

Provision of information may occur as part of training, instruction or supervision, or it may be done independently of these. The type of information to be provided will depend on the types of risks present at the workplace, but, in general, employees and others who may be exposed to risk should be provided with information on issues such as:

- documentation from the risk management process — including risk identification records, risk assessment documentation and risk control records
- the risk control measures in place and how to ensure they are kept in full working order
- operators' manuals
- safe working procedures
- safe use of workplace substances (including information from chemical or product labels and material safety data sheets)
- maintenance and service routines (eg when to request maintenance, and who to ask)
- procedures for repairs (eg tagging and lock-out procedures)

- any special safety information needed (eg safety precautions for working under certain conditions or how to use safety devices, such as emergency stop buttons)
- the types of personal protective equipment needed together with information on fitting, using, cleaning, maintaining and storing equipment
- emergency procedures, and
- where relevant, details on how accidents or incidents have occurred in the past involving the same work activities.

COMMUNICATION, CONSULTATION AND REPORTING

¶6-500 Establishing systems of communication

Merely training employees in safe systems of work will often not be sufficient to ensure workers are able to participate effectively in the risk management process. It is also essential to establish two-way communications systems which enable the benefits of consultation to be realised, and which facilitate optimal reporting with regard to OHS risk management.

Risk management is undertaken for the sake of people who are at risk from the various operations under the control of the employer. The main group of people at risk is employees and so the main focus of risk controls, instruction and training is directed towards them.

A range of other groups and individuals within the organisation will also be interested in the risk management program and its outcomes. These outcomes will need to be communicated to all people who have an interest or are allocated responsibilities. Apart from employees and contractors on site who are likely to be affected by hazards, the outcomes of risk management will need to be communicated to:

- supervisory staff
- the personnel department, and
- record keepers.

There may also be a need to communicate the outcome of the risk management program to the local regulatory authorities and, in some situations, to the emergency services. If a particular hazard could increase the risks to the emergency services, then the nature and extent of the risks involved will need to be communicated to them. For example, the presence of bulk storage of liquid petroleum gas on site should be notified to the fire brigade to ensure that they are ready to act effectively in the event of a fire.

Communicating risk information

The success of the risk management program is determined to a large extent by the effective communication of its conclusions and reporting of risks in order to secure the involvement and commitment of both management and employees.

To successfully communicate risk management information, the following will need to be considered:

- The messages to be communicated — information about the significant findings of the hazard identification, risk assessment and risk control processes must be communicated to the appropriate parties.
- There should be a clear indication of the procedures relating to the reporting of uncontrolled or ineffectively controlled hazards and defects, and the procedures to be followed in the event of serious and imminent danger.
- Any changes in work procedures, work systems and control measures will need to be communicated to those involved. This should involve consulting workers before the changes occur and may involve training and instruction when changes are made. Appropriate supervision will also be needed to ensure that the message has been received and understood, and that it is being used properly.
- The best methods of communication available within the organisation's means should be used — this will require consideration of resources and the current communication channels. Information relating to risk management can be communicated using a range of media from videos to printed material, handbooks and advisory pocket cards. In all cases, accepted organisational channels of communication must be used, with a clear mandate to avoid conflicting or confusing messages.
- The audience must be considered — an evaluation of the intended information recipients should be undertaken to determine the levels of understanding, the existence of language barriers or the use of foreign languages, so that the messages conveyed can be easily understood.
- Follow-up activities will need to be undertaken — all risk management communications should be followed up to check that the intended meaning has been understood and to progress the required action. This may involve people signing for the receipt of risk management information and demonstrating their understanding of its content.
- A system will need to be established to facilitate the issuing of information updates — as reviews of the risk management program are undertaken and changes are made.
- Communication should be a two-way process — risk management is more effective when employees and management, in particular employees and managers who are at risk, are involved in the decision-making process. Communication should not be a system of management disseminating information without having systems in place for listening to and considering feedback from employees. This is vitally important — if employees feel there is a culture of rejecting "bad news" as "negativity", or if they feel they will be blamed or penalised for unwelcome reports about safety hazards, it is likely that negative reports will be suppressed, with a consequent loss of crucial feedback about potential OHS problems. If this happens, the opportunity to address these risks is lost.

Consideration of these factors will assist in ensuring that risk management information is readily communicated.

Barriers to successful communication

Poor communication will result if:

- there is a lack of proper preparation
- the message communicated is unclear or leads people to the wrong conclusion
- wrong assumptions underpin the message
- there has been no evaluation of differing social cultures impacting on communications needs (the most notable being language barriers), and
- there is an organisational culture of "shooting the messenger".

Resolving OHS issues

Good communication procedures are also important if a dispute arises concerning a health and safety matter. The organisation should develop procedures for dealing with issues of this sort.

Employees or contractors should raise the matter with their supervisor, who should attempt to resolve the problem. If the supervisor does not deal with the issue in a way that satisfies the person who raised the concern, they should talk to the relevant health and safety representative or OHS committee member. Typically, this person would approach the relevant supervisor or higher management, if required, about the matter.

If the issue cannot be resolved at this level, the OHS representative should refer the issue to the OHS committee (if there is one) for discussion. The OHS committee should try to resolve the matter before contacting the relevant state or territory OHS authority (usually known as WorkCover or WorkSafe) and making a request for an inspector to visit the workplace to provide advice.

The inspector will firstly determine whether the issue has been addressed appropriately, that is, whether management has been approached regarding the issue, and if it has been referred to the OHS committee, or addressed as required under other agreed arrangements or between OHS representatives and management. If the matter has not been addressed in a satisfactory manner by management, the inspector may decide to attend the workplace or intervene in some other manner.

The inspector's role is to assist employers and employees to achieve a satisfactory outcome where foreseeable risks to health and safety are properly controlled. Other options open to the inspector, such as initiating a prosecution for breach of OHS laws, are covered in Chapter 2. The matter of OHS issue resolution is specifically covered in the legislation is some states. For example, in Victoria, it is the subject of the Occupational Health and Safety (Issue Resolution) Regulations 1999.

Reporting

Good communications systems and a receptive and open management culture will facilitate timely and relevant reporting of OHS matters. This is an essential prerequisite for relevant personnel to take appropriate corrective and preventive action.

Under-reporting breaks the feedback loop to management, if unreported problems are not dealt with. The result of this is that the work may be more hazardous for everyone. Some OHS management practices, such as incentive schemes for sections with the longest run of accident-free days, tend to suppress reporting and are therefore counterproductive. (See "Performance measurement" in Chapter 7 for more discussion of this point.)

Under-reporting of incidents or hazardous situations has been identified as such a significant issue in some companies that systems have been developed to prevent it. For example, work crews are asked to rate each day or each shift on a scale of one to four, with each level having a specific definition based on positive or negative safety behaviour and occurrences.

Notification requirements

There are legislative requirements at the federal, state and territory levels in relation to accident and incident reporting or "notification". These arrangements oblige employers to record the details of certain accidents and incidents to their local OHS authority. The types of accidents and incidents which must be reported are usually:

- accidents resulting in loss of life (which must be reported immediately)
- accidents resulting in a worker taking a number of days off work due to injury, or
- incidents involving damage or potential damage to dangerous items of plant (such as boilers).

Specific forms are produced by the various OHS authorities which must be filled out in the event of an accident or incident. The exact details of the reporting (or "notification") requirements vary between the OHS authorities. In New South Wales, for example, WorkCover NSW has abolished the previous system and, in September 2003, replaced it with a simplified online form which can be completed on WorkCover's website. The principle behind reporting, however, is the same. The primary aim is to compile statistics to assist in determining the causes of work-related injury and disease, so that effective measures may be developed to prevent work-related injury and disease.

RECORD KEEPING

¶6-600 What records to keep and who keeps them

Keeping records related to workplace health and safety is a core OHS function which needs to be monitored by supervisors, line managers and OHS management personnel. Records provide information which may be used to identify OHS problems, to assign priority in addressing these problems and to evaluate progress in bringing the problems under control. Records also yield statistics and other information which enable comparisons to be made, for example, between locations or between one time period and another.

In addition, records are an important element in the relationship between the workplace and the local state or territory OHS authority. Certain records must be forwarded to the authority for inclusion in national OHS data; other records must be kept at the workplace as evidence of the OHS systems in place. One of the first things an OHS inspector will usually do on entering a workplace is to request to see the health and safety records. These provide a first impression of the importance given to OHS in the workplace.

Moreover, record keeping is a fundamental requirement in all OHS legislation. In the worst scenario — if an employer ends up in court for a breach of OHS legislation — records are a very important means of proving the employer's commitment to OHS. Evidence of sincere efforts to efficiently manage workplace OHS can help mitigate any potential penalties for OHS offences.

Types of records

There are many different types of records which need to be kept in relation to workplace health and safety. It is not intended to provide an exhaustive list here, but to give a general overview to assist health and safety managers in identifying the records they may need to keep.

The types of records which need to be collated and maintained include:

- risk management documentation
- incident reports (including reports on accidents, injuries and near misses)
- sickness absence records
- workers' individual histories of exposure (to certain chemicals)
- first aid records
- medical records, including results of medical tests
- results of workplace environmental monitoring (occupational or industrial hygiene monitoring)
- rehabilitation information
- workers compensation records, and
- performance appraisal and training records.

These types of records are kept in compliance with legal requirements and for internal purposes such as planning, budgeting and OHS performance review. Some specific purposes for collecting and analysing this information follow.

Risk management documentation

Records of the risk management process must be kept where the risk assessment indicates a need to control a hazard. There are specific requirements for keeping records in relation to particular types of hazard in the legislation of each state and territory.

An example is to be found in the New South Wales Occupational Health and Safety Regulation 2001, which requires employers to keep records of the risk assessment for hazardous substances if the assessment identifies a risk to health. Where the assessment indicates a need for workplace monitoring or health surveillance, the records must be kept for 30 years, otherwise they must be kept for five years.

Incident reports

Statistical information from incident reports may indicate trends over time and particular areas for concern, such as the involvement of certain:

- work areas and processes
- injury types
- groups of employees, and
- specific locations.

This information can be used to assist in detecting potential problems and in rectifying identified problems. The information, if collected in a standardised way, can also be used to determine the success or failure of chosen interventions — for example, through comparison with information from other equivalent parts of the organisation or previous records.

Sickness absence records

These records relate to individuals and are kept for personnel and payroll reasons. They can also be a valuable tool in occupational health and safety as they may indicate trouble spots, such as unsafe or unhealthy working conditions, or problems with the management of certain areas or people.

Exposure histories

Records of people's exposure to various substances (chemicals) are important in determining the effectiveness of controls. They may also be used for legal purposes, such as in workers compensation cases.

First aid records

Records of first aid treatment should be kept, including the date, time, and a description of the accident or the symptoms of the person seeking treatment. These records should be kept confidential, but they may be used to generate statistical reports for management or health and safety representatives or committees.

Results of medical tests

Results of medical tests may be kept for a number of health and safety reasons including:

- meeting requirements in legislation (such as the need to keep health surveillance records for lead industry workers)
- ensuring that job applicants are fit for the job (or in-company transfer) being sought and that they will not be a danger to themselves, to others, or to the product (such as food)
- providing baseline records with which later comparisons may be made
- alerting medical personnel to any health problems the individual may have in case there is a need to provide treatment on site, or providing information for medical and paramedical specialists to whom the workers may be referred, and
- other reasons related to workers compensation and superannuation/ retirement.

Rehabilitation records

Individual rehabilitation records include a description of the injury or illness, communications between people involved in the rehabilitation program, the roles, duties and deadlines of the parties involved, the return to work plan and the ill or injured individual's progress. These records also show how systems of work have been modified (or need to be) for a reduction of risks to health and safety.

Workers compensation records

Reporting requirements for injury and illness involving workers compensation are provided in the relevant legislation. The compilation of this information contributes to the statistics on workplace health and safety.

Training and performance appraisal records

These records enable a systematic approach to training and performance appraisal. People should be trained in safe work practices and in control measures for the particular hazards associated with their jobs. Induction training plus ongoing training or skills updating should be audited for each person. Performance appraisal should include reference to health and safety commitments.

There are legal requirements for keeping certain types of training records related to OHS. An example can be found in the New South Wales Occupational Health and Safety Regulation 2001 which requires employers to provide appropriate training for employees working with hazardous substances; records must be kept of the training carried out.

Record keeping standard

The National Occupational Health and Safety Commission has published a national standard on record keeping. This document, AS 1885.1: *Workplace injury and disease recording standard*, aims to provide individual workplaces with guidance on establishing a simple and inexpensive method of recording information on work injury and illness. This standard, which is useful for both large and small organisations, has been adopted in most jurisdictions and is a guide to good practice.

Chapter 7

MEASUREMENT, EVALUATION AND REVIEW

¶7-100 Introduction

The purpose of this chapter is to examine some considerations regarding the measurement, evaluation and review of occupational health and safety (OHS) performance. Also covered are techniques commonly used in accident investigation.

Successful health and safety performance cannot be assumed merely because an organisation has implemented a particular program. Like every other aspect of the organisation's performance, each health and safety program must be evaluated by management in order to determine whether it has been successful and whether any improvements can be made. Without the necessary resources devoted to evaluation, the organisation is committed to "blind acceptance" of the program originally chosen.

Evaluating health and safety performance has the following elements:

- consulting workplace stakeholders about the purpose and process of evaluation
- choosing measurable objectives
- measuring what has happened
- comparing these measurements against the original objectives, and
- deciding on new or revised plans.

METHODS OF MEASUREMENT

¶7-200 Selection of method

Methods of measuring the outcome of a safety program will depend on the program. For example, a hearing conservation program can be assessed by taking measurements of noise levels and by audiometric testing of employees. Safety auditing, a form of ongoing health and safety measurement, is discussed at ¶7-300 et seq. (See also ¶7-600.)

It was noted in ¶5-300 that objectives have to be formulated so that the degree to which they have been achieved can be measured by suitable indicators. If objectives are too vague, or not expressed in a way that facilitates subsequent measurement, the opportunity to assess the organisation's progress in achieving its aims is lost.

¶7-220 Monitoring

To effectively monitor the organisation's progress in meeting its OHS objectives, it may be necessary to make specific regular measurements, inspections or other checks. Examples include:

- checking to ensure controls are effective in relation to potentially hazardous processes
- checking whether legislation, standards or industry practice has changed in ways that have implications for existing risk controls

- checking to ensure that potentially hazardous plant, such as pressure vessels, complies with legal requirements, and
- checking that basic health and safety rules are followed.

With some hazards, such as noise or hazardous substances, it may be necessary to undertake environmental testing, eg for noise levels or the presence of atmospheric contaminants (such as dust or chemical vapours).

Measurements of this sort require expertise and equipment which OHS managers, health and safety representatives and other workplace parties will not normally possess. For example, the measurement of noise levels presents an array of technical challenges including differing levels of noise at different stages of work activities and in different parts of the workplace, the varying duration of exposure of individual workers, and how to factor intermittent noise into the assessment.

It is therefore essential for workplace personnel to recognise the limits of their own expertise and seek expert advice as appropriate. See Appendix C for the types of expertise available from the various categories of OHS professionals, such as occupational hygienists.

Equipment used by occupational hygienists to measure noise or atmospheric contaminants needs to be selected according to the types of agent or condition to be measured, the area where the measurements are to be made, the work activities being carried out and the level of risk. The equipment must be correctly maintained and calibrated, with the appropriate scale selected. A sampling process must be defined and arrangements made with the workplace to collect the required data, and advise those involved of any requirements.

Employees and others affected by the monitoring process should be consulted and informed of the purpose of the exercise. The operability of the equipment should be checked, and arrangements made to ensure the health and safety of people undertaking monitoring activities.

After readings have been taken, any special equipment used should be dismantled, cleaned, stored or disposed of as appropriate. The results must be interpreted and evaluated against a recognised standard and legislative requirements. For example, concentrations of hazardous substances in the air workers are breathing should not be higher than the recommended maximum exposure level.

The results must be documented and communicated as necessary to management and workers. Records should be kept as required by legislation. For example, in the case of measurements made to assess risks of exposure to hazardous substances, New South Wales legislation requires the records to be kept for 30 years.

Other forms of measurement which might be used by an OHS consultant include personal dosimetry, which is a method of measuring the exposure of a particular individual to a given hazard such as noise.

¶7-230 Health surveillance

Health surveillance is the monitoring of individuals for the purpose of identifying changes in health status that may be due to occupational exposure to a hazard.

Health surveillance is prescribed by legislation in relation to some hazards, such as lead. This may take the form of biological testing, eg hearing testing or checking blood or urine for the presence of heavy metals, or metabolites of pesticides.

¶7-240 Documentation and records

Procedures for monitoring and measurement should be documented, and their effectiveness evaluated. Equipment used for the purpose of monitoring and measurement should be properly calibrated, maintained and stored.

Records of monitoring and measurement should be kept as required by legislation. For example, in New South Wales employers must keep records of risk assessments indicating a need for health surveillance, and the results of any health surveillance, for at least 30 years after the date of the last entry in them. Such records are also a valuable information resource for the organisation to check whether its risk control strategies are achieving their aims.

Results and records should be retained and stored in a format that enables them to be readily retrieved in accordance with regulatory requirements and standards.

¶7-250 Employees' right to know

Employees at a workplace are entitled to be consulted about and know the results of any evaluation of a health and safety program in which they have participated. A joint effort by management and employees to improve health and safety should involve all parties knowing of any programs that are introduced and the results of those programs. Note that legislation generally provides health and safety representatives or committee members with the right to all safety information concerning the workplace — see ¶2-140.

It may be worthwhile for management to call a meeting of all employees involved in a particular program after an evaluation of that program has been completed. The results of the evaluation can be presented, and options for amending the program discussed. Where a large number of employees are involved (such as a program that affects the whole workplace), this process could be carried out through the health and safety representatives and/or committee. All staff should be informed by circular or a message on safety noticeboards.

HEALTH AND SAFETY AUDITING

¶7-300 Overview

Health and safety auditing has already been referred to in Chapter 6, where it was described as a systematic and periodic review of the whole OHS management system, including the policy and programs used to promote OHS and to prevent workplace accidents and incidents. All organisations, whether large or small, can benefit from applying safety auditing techniques. Audits may be carried out over the whole organisation, or over sections of the organisation, or in relation to a particular aspect of risk control such as electrical safety or fire safety.

Occupational health and safety auditing involves evaluating health and safety performance using checklists and measurable criteria. It aims to determine whether the health and safety management system as a whole (the organisation's policy, objectives and targets, management program and operational controls) has been properly implemented, whether all the required elements are in place and whether the organisation is meeting both the objectives set in its OHS policy and legislative requirements. If the audit reveals areas of weakness, either uncontrolled risks, areas where changes will affect risks, or inoperative elements of the management system, these should be addressed.

The two essential features of a safety auditing system are that the audits be repeated at intervals and that the information collected be used.

¶7-320 Conducting an audit

Each safety audit should build upon previous audits. Hazards that are identified by a previous audit can be further investigated by the next. This may involve fine-tuning the checklist that relates to that particular area. The results of the initial process of hazard identification discussed in Chapter 6 may be used as a standard to plan an auditing program.

If auditing over time shows a marked improvement in a particular area, it may be possible to increase the time between audits of that area.

Checklists

Checklists are the traditional method of safety auditing. Checklists allow a comprehensive approach to auditing (providing the checklist is well prepared). They also result in a standard approach to the audit, so that it can be compared with past audits.

However, checklists do have the drawback of stifling thought and discussion. Auditors should not limit themselves to the list, but should test and ask questions if they come across potential hazards.

Remember that, wherever possible, questions on checklists should be framed so that answers are given as measurable standards, rather than opinions or yes/no answers.

"What if?" method of auditing

One useful approach is to use a "what if?" method, that is, auditors ask themselves what would happen in certain hazardous situations that may occur. For example, the auditor may ask "what if this machine leaked oil on to the floor?" or "what if personnel from another department were present while a chemical shipment was being unloaded?". This method can be used to supplement checklists.

Who conducts the audit?

Safety auditing is best performed by small groups of (usually) three people. It may be useful to include an employee from the particular area being audited (possibly a health and safety representative). Some auditing topics (such as noise, chemical safety and ventilation) will require an expert on that topic to be part of the team.

Another option is to engage an external consultant to conduct the audit. This may be necessary in areas that require expert knowledge. In other areas, it may be desirable to engage a consultant experienced in safety auditing to perform an initial audit, which can then be used as a base for future audits.

¶7-330 Scope of the audit

Health and safety audits can cover as many aspects of a company's operation as those responsible for the audit can think of. The survey at ¶4-150 can be used as the basis for developing an audit or, if a briefer or more basic format is preferred, the following starting list provides some examples of the areas that can be dealt with and the questions that can be asked. An extensive "paper system" does not necessarily result in better OHS results.

This list should be expanded to suit the particular organisation. Some of the more specific areas at the end of the list may not apply to all organisations. Note also that checklists can be restrictive if relied on too rigidly.

Starting list for health and safety auditing

Health and safety policy

- Does a policy exist?
- To what extent are employees aware of it?
- Is it explained at induction?

Health and safety representatives/committees

- How much training do they receive?
- How active are they?
- How often do they meet with management?
- Which suggestions have resulted in useful changes?

Training

- Is health and safety training provided as part of induction?
- Are all employees aware of the health and safety requirements of their position?

- Have all workers received sufficient training to carry out their duties safely and competently?

Risk management

- Have hazards been identified, and risks assessed?
- Have risks been controlled in accordance with the hierarchy of risk control?

First aid

- What first aid facilities are available?
- Do they comply with legislative requirements?

Incident investigation

- What is the system for accident/incident investigation?
- What follow-up action has resulted from incident investigation?
- Have unsafe situations been remedied so that incidents will not recur?

Incident records and statistics

- Is there an incident recording system?
- What statistics are compiled?
- What trends do the statistics demonstrate?
- Is any use made of the information collected?
- Is the organisation's safety record improving?

Workplace inspections

- How often are inspections conducted?
- How is information from inspections used?

Housekeeping practices

- Are there instances of rubbish being left at the workplace?
- Are slip or trip hazards present?

Lighting

- Is lighting suitable for the tasks performed?

Ventilation

- Are there build-ups of atmospheric contaminants?
- How often is air-conditioning equipment inspected and cleaned?
- How effective is air movement?

Noise levels

- Has employee hearing deteriorated (as measured by audiometric testing)?
- What levels of noise exposure do employees experience?

Protective equipment

- What protective equipment is used?
- Is it possible to make use of solutions further up the hierarchy of risk controls to remove the need for this protective equipment?
- How often is the protective equipment actually used?
- Is the equipment properly maintained?
- Are employees trained in the proper fitting of personal protective equipment?

- Are sunscreens, hats and shades provided for outdoor workers?

Fire safety

- What firefighting equipment is available?
- Is a firefighting team trained?
- How often are fire drills conducted?

Emergency plans

- Is a plan in existence?
- Are the employees aware of the plan?
- Have regular evacuation drills been held?
- Have the emergency authorities been informed?

Injury management program

- Has the program reduced the costs of injuries?
- Are the injured employees satisfied with the services provided?

Chemical safety

- What chemicals or other hazardous substances are kept or generated at the workplace?
- Are MSDSs in existence and readily available?
- Do employees work in accordance with the precautions listed on the MSDS?
- Is labelling adequate?
- Are the atmospheric levels of the hazardous substances in use lower than the relevant exposure standards?

Manual handling

- Have manual handling risks been assessed and satisfactorily controlled?
- Are relevant employees trained in proper manual handling techniques?
- What instances of back injuries or other musculoskeletal disorders occurred?
- Is suitable materials handling equipment provided?

Plant safety

- Are proper maintenance routines in place (what is the frequency of inspections)?
- Are machine guards provided?

Computer workstations

- Are workstations designed in accordance with ergonomic principles?
- Have any instances of occupational overuse syndrome been reported?
- Do workers have adequate rest breaks?

Electrical safety

- Does all electrical installation comply with the relevant Australian standards?
- Is equipment inspected/tested and certified in accordance with Australian standards?
- Are all employees working with electrical equipment adequately trained (do unsafe occurrences suggest otherwise)?

ACCIDENT/INCIDENT INVESTIGATION, REPORTING AND RECORDING

¶7-400 Introduction

An essential part of managing health and safety risks is to investigate accidents and near-miss incidents, so that the causes can be addressed and the chances of the same or similar incidents recurring can be eliminated or at least reduced. This is another way in which hazards can be identified and risks assessed and controlled, and it requires both a systematic means of investigation and the maintenance of a record system to monitor progress.

¶7-410 What are accidents and incidents?

As defined at ¶1-200, a workplace accident may be described as any event arising out of employment that results in human injury or damage to property, or the possibility of such injury or damage.

For the purposes of accident prevention, however, it is also important to record and investigate all "near-miss" situations (that is, events which did not result in injury or damage but that had the potential to do so). The term "incident" has been more widely used in recent years and is tending to replace "accident" because it conveys a wider scope of events with potential OHS implications.

¶7-420 Accident causes

The complex nature of accidents/incidents has already been referred to in Chapter 3 at ¶3-130. All accidents or near-miss situations represent a failure in, or an inadequate, OHS management system at the workplace.

Identifying the real cause of an incident is a complicated process. Investigation may reveal several coincidental causes that create a chain of causation factors, none of which would have resulted in an accident by themselves.

The basic concept of accident prevention is that incidents can have several causes, each of which can be identified and controlled.

For example, an unsafe act may have been due to the lack of a defined company policy, or the lack of proper training, or faulty supervision, leading to a breach of rules. Alternatively, it may have been due to risk taking or skylarking. Another cause of accidents could be cost cutting, such as lack of adequate maintenance or use of cheap, unsatisfactory protective equipment.

The following examples indicate the complexity at times of the process of identifying causes.

1. An employee slips on an oil patch in a warehouse

Is it because of:

(a) bad housekeeping (that is, no one wiped up the spilt oil, which could be due to inadequate information, training or supervision; or the absence of documented safe working procedures)

(b) poor maintenance (if the oil was dropped by a forklift truck), and/or

(c) faulty management (if the transport officer is on leave and no one renewed the maintenance contract)?

2. A truck driver ignores signs of brake failure and is injured

When a truck driver ignores a sign of brake failure and is injured when the truck fails to stop at an intersection and collides with a car, the accident is due to the driver's failure to respond to the danger signal. One should therefore look for reasons why the driver chose to ignore it, instead of dwelling solely upon the accident itself. Was it because:

(a) the delivery schedule was too tight

(b) the driver was over-confident or inadequately-trained (eg driver thought the sign of brake failure was within tolerable limits), and/or

(c) it was the last load for the day?

The implication of these examples is the need to focus on all the links in the chain of events that led to the unsafe acts or conditions, rather than the consequences (that is, the injury or damage).

Among the most common underlying causes are the following general categories:

- management's neglect of its obligations to provide a safe system of work, a safe place of work, competent staff, a sufficient number of workers and proper plant and equipment
- management's failure to foresee a risk which should have been obvious
- lack of attention to health and safety procedures and practices (such as failure to assess manual handling risks or failure to replace a guard)
- inappropriate or poorly maintained personal protective equipment (PPE), or workers' failure to use PPE supplied
- inadequate training, instruction and supervision of employees (or failure to understand instruction), even at the original induction stage of employment
- inadequate attention to maintenance and "housekeeping", and
- failure to provide adequate safeguards and/or suitable equipment for potentially dangerous situations.

Attention should also be paid to psychological factors. For example, if an employee finds a job boring, monotonous, or well below his/her capabilities, the result may be alienation from the job, emotional stress, lack of care and loss of concentration, all of which increase the chances of accidents occurring. The solution to these problems will involve an examination of job design and the match of employees' capabilities to suitable jobs.

Job demands must also be considered, such as the effects of long working hours, shiftwork, deadlines or other time pressures, fatigue, piecework schemes which emphasise increased production at all cost, and other physical and mental stresses.

¶7-430 "Accident proneness" and blame

Current approaches to occupational health and safety avoid making value judgments as to whose "fault" caused an accident or who can be "blamed". This is reflected by no-fault workers compensation systems. Instead, the system aims at investigating and explaining accidents with a view to identifying the fault in the work system that led to the accident, and identifying control measures by which accidents can be avoided.

In the same way, the concept of identifying certain people as being "accident prone" has been rejected by social scientists due to the lack of evidence of any specific personality characteristic that makes a person susceptible to accidents. Focusing on only one cause or person tends to obscure the underlying causes that may have contributed to the accident.

It is possible, however, for employees to become more prone to accidents in time because of undetected, normal deterioration of eyesight, hearing, body movements, and/or the use of medication. It is also essential to be alert for "repeaters" in relation to type of injury or time of year, such as recurring back problems, heatstroke or unsuitable training of seasonal workers.

¶7-440 Common types of incidents

Workers compensation statistics suggest that the following are the most common types of accidents:

- manual handling injuries/over-exertion or over-strenuous movements
- stepping on, striking against or being struck by objects, and
- slips, trips, falls and jumps by persons.

Each of these usually has a variety of causes that can be addressed by risk management, effective health and safety training practices and good supervision.

¶7-450 The steps of incident investigation

As with other management activities, incident investigation requires certain defined steps to be taken by nominated people in terms of procedures, use of resources, paperwork and follow-up action. These steps, listed below, are discussed in the remainder of this chapter:

- what should be investigated (¶7-460)
- who should investigate (¶7-470)
- investigating incident causes (¶7-480)
- special equipment (¶7-490)
- incident recording systems (¶7-500)
- incident reporting requirements (¶7-510)
- statistics (¶7-520)
- report writing (¶7-530)
- preventing recurrence of incidents (¶7-540)

- feedback to employees (¶7-550), and
- disciplinary policy (¶7-560).

¶7-460 What should be investigated?

All accidents or near-miss situations should be investigated. This investigation should take place as soon as possible after the incident occurs. Promptness of investigation is important as crucial evidence can be disturbed or destroyed with the passage of time. A camera is a useful tool as it captures conditions as they existed at the time of the accident. After a serious incident, in general, the accident site must not be disturbed. That is, any plant or equipment, substance or thing at the site should not be used, moved or interfered with after it has been involved in a serious incident, and the surrounding area at the accident site must not be disturbed. A serious incident here means one involving a fatality or other serious consequences.

Naturally, these non-disturbance requirements should not prevent any action aiming to help or remove a trapped or injured person or to remove a body; nor should they prevent anything being done to avoid injury to a person or damage to property, or the taking of any essential action to make the scene safe or to prevent a further occurrence of an incident.

Both a survey of the accident location and a chronology of the accident events should be obtained. This will involve accurate measurement and probably the preparation of a sketch map. As many eyewitnesses as possible should be questioned. The observations made by different witnesses will probably conflict so it is better to concentrate on collecting witnesses' reports on the spot and reconstructing the probable chronology after the initial questioning.

Accident investigation should not be confined to on-site analysis. Statistics and past trends may indicate particular areas for concern involving certain work processes, injury types, time losses, groups of employees, or locations. Details of the investigation should be conveyed to similar work situations within the organisation to promote awareness of possible undetected hazards there.

¶7-470 Who should investigate?

Each accident or near-miss should be investigated by the immediate supervisor of the section concerned, although assistance may be sought from others. Selected incidents may be separately investigated by the health and safety officer/manager, occupational health nurse, and health and safety representatives and/or committee. Note that anyone who carries out an investigation should have the training to do so.

The following is a list of desirable skills that the persons in charge of the investigation process should ideally possess:

- a range of basic investigative skills, recording skills (photography, drawing of plans) and technical knowledge (of work processes, equipment and work layout)

- ability to interview people, such as the injured person, witnesses and others with relevant knowledge
- a capacity to understand the needs and reactions of employees
- skill in report writing
- ability to observe and note details (for example, when attending the accident site), and
- ability to recommend preventive action, implement it where it is within the person's authority to do so, and assess the effectiveness of such action.

It is advisable that more than one person be present at the on-site accident investigation. Often an outside authoritative source may need to be called in, immediately if possible, or at a later date when the work task or process is in progress again.

¶7-480 Investigating accident causes

Carrying out an investigation aimed at identifying the underlying causes of an accident or near miss will require asking some basic questions. The importance of concentrating upon the underlying cause, rather than the outcome, was emphasised at ¶7-420. The implications are that it will usually not be sufficient to interview the employee(s) directly involved. In addition, an examination of the workplace and its operation methods will be required, as well as the collection of background information.

If a person is injured carrying out a particular task, it is necessary to establish whether he or she considered alternative ways of doing that task. This is important because if the person knew no safer way of doing the task, then the most appropriate form of remedial action may be training or instruction. If, however, the person did know of a safer way and chose, for whatever reason, the more dangerous way, then training is unlikely to be an effective type of remedial action and other means of addressing the causes will need to be considered.

One useful question to ask employees during an interview is something along the lines of: "What do you suggest to prevent a future recurrence of this accident/incident?" This may be an appropriate subject for review by a workplace committee.

> The primary objective of accident investigation from an organisational and OHS authority perspective is to prevent future accidents, not to assign blame. This should be stressed to employees who are interviewed following an accident, to ensure that they provide all relevant information. Government authorities, however, may need to investigate incidents for purposes of initiating prosecution.

¶7-490 Special equipment

Equipment required for investigating accidents or incidents will of course vary according to the type of workplace, but the following table suggests a basic range of equipment that could be useful. A kit should contain measuring, recording and marking devices, with each item readily accessible and ready for immediate use.

Camera and film (preferably VHS with playback, digital or polaroid to allow for immediate assessment of the adequacy of the photo).

Cassette tape recorder for witness interviews, oral notes, observations and reports.

Sound level meter to measure and analyse noise levels.

Gas and vapour analyser to measure carbon monoxide and other gas vapour concentrations.

Electrical receptacle tension tester to test electrical continuity and grounding of receptacles and distribution cords.

Wire/hemp rope calculators to measure rope size for analysis of load capability of ropes used for lifting and stabilising.

Callipers for precision measurement of parts: inside and outside dimension callipers as pertinent.

Sample containers of type, material and size appropriate for collection of fuels, lubricants, coolants and other substances.

Protective equipment: a range for use by investigators.

Note that some of this equipment, such as sound level meters or gas and vapour analysers, should only be used by OHS professionals with the relevant training (see ¶7-220 for more details on the use of special equipment).

¶7-500 Incident recording systems

All employers should establish a system for recording workplace injuries, near misses and diseases. The availability of reliable and usable data is essential for the development, monitoring and evaluation of health and safety strategies.

Information from an accident recording system can be used to complete statutory accident notification requirements (see ¶7-510), but its prime function is to initiate preventive action to control future incidents. With this in mind, any recording system must allow the identification of trends in the types of injuries, time of occurrences, location of occurrences and agents of injuries or diseases. The amount of time lost as a result of each injury/disease should also be recorded. Recording this information allows safety statistics to be compiled (see ¶7-520).

The National Occupational Health and Safety Commission (NOHSC), in conjunction with Standards Australia, has developed the *National Standard for workplace injury and disease recording* (AS 1885.1-1990). The standard provides employers with a simple method of recording comparable data covering accidents, near-misses and diseases at the workplace. It can be used by both large and small employers. The sample recording form, which is published as an appendix to the standard, is reproduced on the following pages. Note that it is not necessary to use the sample form in order to implement the standard. Some organisations may wish to design a form with additional information.

Medium and large organisations may want to computerise their workplace accident and disease records. Standards Australia offers a software package that allows users to record all necessary information on workplace injuries in a form that fully complies with AS 1885.1.

Use of AS 1885.1 will satisfy requirements contained in state health and safety legislation regarding the recording of injuries. An example of a Workplace Injury and Disease Recording Form is on the following pages.

AS 1885.1—1990

WORKPLACE INJURY AND DISEASE RECORDING FORM

Reference number [ ]

Personal details of the injured worker

1. Surname [ ]
   Given names [ ]
2. Sex (M or F) [ ]
3. Date of birth (Day / Month / Year) [/ /]

Basis of employment

4. Starting time:
   1 [ ] 0600 – 1159 hours
   2 [ ] 1200 – 1559 hours
   3 [ ] 1600 – 0559 hours
5. Shift arrangement:
   1 [ ] Fixed, standard or flexible hours
   2 [ ] Rotating shift
6. Number of hours:
   1 [ ] 8 hours or less
   2 [ ] more than 8 hours (excluding overtime)

Job details

7. Description of occupation or job title [ ]
8. Main tasks performed [ ]
9. Training provided:
   1 [ ] Induction training
   2 [ ] Task specific training
   3 [ ] Both of the above
   4 [ ] Neither of the above

Details of the injury or disease

10. Date injury occurred or disease reported (Day / Month / Year) [/ /]
11. Time injury occurred [ ] (24 hour clock format)
12. Nature of injury or disease [ ] Code [ ]
13. Bodily location of injury or disease [ ] Code [ ]

14. Description of occurrence of injury or disease:
   - In which part of the workplace did the injury or disease exposure occur?
     (e.g. machine shop, freezer room, No. 2 mine) [ ]
   - What was the worker doing at the time?
     (e.g. driving a fork lift truck, lifting bags of cement, typing) [ ]
   - What happened unexpectedly?
     Include the name of any particular chemical, product, process or equipment involved.
     (e.g. brakes failed on fork lift truck, slipped on wet floor, scaffolding collapsed, arm started hurting while typing on a word processor) [ ]
   - How exactly was the injury or disease sustained?
     Include the name of any chemical, product, process or equipment involved.
     (e.g. hit head on cabin of fork lift truck, lacerated knee when landing on ground, arm hurt after long period of typing) [ ]

(See pages 25 – 28 of the Standard)

	Code
Mechanism of injury	[ ]
Breakdown agency	[ ]
Agency of injury	[ ]

AS 1885.1—1990

Lost-time injury/disease

Additional questions to be answered for cases which result in a fatality or permanent disability, or where there was time lost from work of one or more days/shifts. These questions should be completed as soon as possible after the injury or disease is reported.

15. Employee's preferred language [ ]

16. Type of employment:
- 1 ☐ Full-time permanent
- 2 ☐ Full-time casual
- 3 ☐ Part-time permanent
- 4 ☐ Part-time casual

17. Type of employee:

Wage/salary earner:
- 11 ☐ Trainee
- 12 ☐ Outworker
- 13 ☐ Apprentice
- 14 ☐ Pieceworker (other than Outworkers)
- 15 ☐ Other

(Note: most employees will fall into this category) →

Self-employed: 20 ☐ (including contractors and sub-contractors)

Unpaid worker:
- 31 ☐ Work experience
- 32 ☐ Other

18. Worker's experience in task being carried out when injury or disease occurred — Years [ ][ ] Months [ ][ ]

19. Proportion of shift worked:
- 1 ☐ 25% or less
- 2 ☐ 26% – 50%
- 3 ☐ 51% – 75%
- 4 ☐ 76% – 100%
- 5 ☐ Overtime

Details of person completing this form

Name [ ]

Position [ ]

Signature [ ]

Date [/ /]

Outcome of injury/disease

Questions 20 – 24 are about information that is not available at the time of the report of injury or disease. These questions should be answered as soon as the information becomes available. For some occurrences, such as where there was no time lost, some of these questions will not be relevant.

20. Rehabilitation:
- 1 ☐ Required → Date of commencement of rehabilitation program — Day Month Year [/ /]
- 2 ☐ Not required

21. Was the injury or disease:
- 1 ☐ Fatal
- 2 ☐ Non-fatal

22. Preventive action proposed or taken:

[ ]

(Tick one or more boxes as appropriate)

	Proposed	Taken
Change to induction training	11 ☐	12 ☐
Change to ongoing training	21 ☐	22 ☐
Equipment/machinery modifications	31 ☐	32 ☐
Change to work procedures	41 ☐	42 ☐
Change to work environment	51 ☐	52 ☐
Equipment/machinery maintenance	61 ☐	62 ☐
Other job redesign	71 ☐	72 ☐
Other preventive action	81 ☐	82 ☐

23. Date of resumption of work on:

	Day / Month / Year
Short-term alternative duties	/ /
Permanent alternative duties	/ /
Normal duties	/ /

(Enter each date when applicable)

24. Total number of working days lost [ ][ ][ ][ ]

(Should be completed only when the worker has resumed permanent duties)

¶7-510 Incident reporting requirements

State workers compensation legislation generally requires employers to keep a register of injuries, and in some cases to report injuries to the relevant authority. The *National Standard for workplace injury and disease recording* referred to in ¶7-500 may not satisfy this requirement, as the workers compensation legislation may prescribe a specific form of register.

In addition, as mentioned at ¶6-500, occupational health and safety legislation requires that certain types of accidents, usually those involving death or absence from work beyond a certain number of days, must be reported to the body that administers health and safety in that state.

In Victoria, for example, the *Occupational Health and Safety Act 2004* requires employers or self-employed persons to notify WorkSafe Victoria as soon as they become aware of an incident that results in death or serious injury. Serious injuries include, but are not limited to, those that require immediate medical treatment, or medical treatment within 48 hours of exposure to a substance.

Incidents must also be reported to WorkSafe if they expose anyone to an immediate health or safety risk, including:

- the collapse, overturning, failure or malfunction of, or damage to plant that is required to be licensed or registered
- the collapse or failure of an excavation or of any shoring supporting excavation
- the collapse or partial collapse of a building or structure
- an implosion, explosion or a fire
- the escape, spillage or leakage of any substance, including dangerous goods, or
- the fall or release from a height of any plant, substance or object.

Within 48 hours of becoming aware of such an incident, the employer or self-employed person must give the authority a written record of what occurred. They must keep a copy of this record for at least five years and make copies available to:

- people injured or exposed to risk by the incident, and their representatives
- the representatives of anyone who died as a result of the incident, and
- the HSR for the designated workgroup and the members of any health and safety committee established.

¶7-520 Statistics

Australian Standard AS 1885.1 (see ¶7-500) sets out the following statistical formulae for use in industry.

Incidence Rate: the number of occurrences of injury/disease for each one hundred workers employed.

$$\frac{\text{number of occurrences in the period}}{\text{number of workers}} \times 100$$

The "number of occurrences in the period" refers to all injuries/diseases that resulted in lost time of one day/shift or more during the specified period.

The "number of workers" is defined as the average number of workers who worked in the recording period. Persons who were absent from work on paid or unpaid leave for the entire period should be excluded from this calculation.

Frequency Rate: the number of occurrences of injury or disease for each one million hours worked.

$$\frac{\text{number of occurrences in the period}}{\text{number of hours worked in the period}} \times 1{,}000{,}000$$

The "number of occurrences in the period" is defined in the same way as for the incidence rate.

The "number of hours worked in the period" refers to the total number of hours worked by all workers in the recording unit including, for example, overtime and extra shifts.

Average Time Lost Rate: the average time lost per occurrence of injury/disease. This rate provides a measure of the severity of the occurrences being experienced by workplaces over time.

$$\frac{\text{number of working days lost}}{\text{number of occurrences in the period}}$$

The "number of occurrences in the period" is defined in the same way as for the incidence rate.

The "number of working days lost" refers to the total number of working days, irrespective of the number of hours that would normally have been worked each day, that were lost as a result of the injury/disease up to a maximum of 12 months for any individual occurrence. For the purposes of calculating the average time lost rate, occurrences that result in a fatality should be assigned a time lost of 12 months (220 standard working days).

In industries where injury or illness is not uncommon, it is recommended that medium and large organisations prepare these statistics every month, while smaller organisations may only need to calculate rates on a six-monthly or annual basis.

The main function of these statistics is to measure trends over time so that the organisation can tell whether its record is improving, stable or deteriorating. For this reason, it is important that data be retained for several years. The statistics also provide a guide to the effectiveness of any corrective action taken as part of a health and safety program.

Converting information to chart form, particularly monthly figures over a 12-month period, can be useful in highlighting periods in which more accidents/injuries occur than in others. It may also be worthwhile to use a chart to compare accident/injury rates with absenteeism statistics, to check for similarities between the two.

In addition to these basic statistics, there are other calculations that might prove very useful in identifying particular "trouble spots" within an organisation. For example, accident statistics could be isolated and classified by department, type of work process, agent of injury/disease, age group of employees, length of service, length of time off work, type of injury and other relevant classifications. One advantage of such an approach is that "good" and "bad" sections of an organisation can be compared and, if relevant factors can be found, it may sometimes be possible to transfer the features of the "good" sections to other sections.

Caution on the use of statistics

It is important, however, that statistics be used with caution. Comparisons with other organisations will often be invalid, due to different work processes, employee backgrounds and so on. They could be difficult to obtain, as many firms would be reluctant to disclose such information.

For the purpose of communicating with employees, a simpler form of statistical presentation may be more effective. For example, many organisations use a notice setting out achievable "targets" such as a 5% injury reduction each year.

It is essential to ensure that, if the above types of notices or targets are used, they must not result in suppression of accident reporting. It has often been noted that such notices can create a social pressure that leads to under-reporting, and is therefore counterproductive. See also ¶7-600 for discussion of this point.

¶7-530 Report writing

It may be desirable to provide health and safety staff and all managers and supervisors with some training in report writing. Such training will help them provide relevant information and express themselves clearly as outlined below.

First, to convince management that there is a problem and to spend money on the remedial action proposed, it will be necessary to present facts in a clear and logical fashion and demonstrate how the proposed solution will be the most effective or most cost-effective. Ideally, alternative solutions should also be suggested where feasible.

Second, a report can provide a check on the adequacy of incident investigations, raising questions as to whether additional inquiries should have been made, or whether there is enough information to form definite conclusions. Further investigation may be necessary after this feedback before remedial action can be decided on.

Finally, the degree of loss or potential loss incurred by an incident should be quantified in monetary terms. A simple costing of the fundamental areas affected will provide management with an idea of the cost involved. See ¶1-140 for discussion of direct and indirect costs of accidents.

¶7-540 Preventing recurrence of incidents

It must be remembered that the main purpose of investigating incidents is to prevent their recurrence. It is possible that an accident or incident will reveal the existence of hazards or shortcomings in the management system that were not previously identified by inspections, the safety survey or other efforts to ensure all risks were identified. If this is the case, these new insights should feed directly into the risk management system described in Chapter 3.

¶7-550 Feedback to employees

A successful health and safety program will require employee awareness of issues, as well as their support and motivation both to work safely and to assist with the reporting and investigation of unsafe situations. This commitment will only be achieved if employees and their representatives can be convinced that management is acting in the employees' interests. This in turn will require constant feedback to employees on incident investigations and any changes that result from an investigation.

Information that could be supplied to employees includes the following:

- the physical results and costs of incidents to both employer and employees
- the causes of an incident, giving both employer and employee versions if these differ
- progress reports on any long-term projects and changes being undertaken, and
- statistical information on a regular basis, perhaps in chart or graph form for ease of understanding.

Management will need to follow up on the progress of corrective action as well. The timing and responsibility for doing this could be established from the incident report or from other records. Follow-up actions on the health and safety audit or surveys should be conducted periodically and not treated as "one-off" exercises.

¶7-560 Disciplinary policy

There are times when health and safety matters will require the disciplining of employees. This is a serious matter that should not be undertaken merely so that someone can be seen to be taking the blame for an incident.

Employees have legal responsibilities under OHS legislation and, if an employer condones an employee's breaches under that legislation, the employer is in effect in breach of the legislation as well.

A disciplinary policy should aim to alter the employee's behaviour in a positive way, rather than to "punish" the employee. The policy needs to be consistent and have the cooperation of employees. Factors to consider in the policy include actual and potential money costs, actual and potential injuries, related circumstances (such as equipment and working conditions), concealment by anyone of the occurrence, the employee's past record and treatment of any similar previous cases.

Other relevant factors include the employee's contract of employment and the organisation's rules on other matters.

PERFORMANCE MEASUREMENT

¶7-600 Ways of measuring performance

As noted in Chapter 5 at ¶5-300, performance indicators are a means of measuring whether OHS objectives and targets have been met, and highlighting areas where improvement is needed. While negative outcome measures such as the lost-time injury frequency rate (LTIFR) have been the standard types of indicators commonly used in the past, it is recommended that positive performance indicators be used to supplement traditional measures.

In recent years many Australian organisations have adopted the aim of achieving best practice, or world class performance. This has required the introduction of sophisticated approaches to performance measurement so that organisations can evaluate their progress both in relation to other organisations and their own past performance. Performance monitoring can be applied to numerous areas within an organisation, including production, finance or costs, environmental matters and the health and safety of workers.

Performance indicators have been defined as "a statistic or other unit of information which reflects directly or indirectly the extent to which an anticipated outcome is achieved, or the quality of the processes intended to achieve that outcome". As previously discussed, performance indicators can either be quantitative or qualitative, but there are also other aspects to consider.

Performance indicators are often characterised in terms of whether they are reactive (that is, reacting to events such as accidents or dangerous occurrences) or proactive (that is, they attempt to assess progress in the organisation's efforts to prevent accidents or incidents, or mitigate the worst consequences of such events). A distinction is also often drawn between objective and subjective measures.

The most important feature of performance indicators is that they should provide management with feedback which can be used to improve the management of OHS, and in this way contribute to the control of health and safety risks.

Traditional approaches — "outcome" measures

The traditional approach to the measurement of OHS performance has focused on negative outcomes such as fatalities, workers compensation cases or lost-time injuries. Other negative outcomes sometimes used as indicators include sickness absences, complaints, criticisms, downtime, spoiled work and regulatory agency enforcement actions. The most common outcome indicator is the LTIFR which has long been used as the standard measurement of OHS performance.

Such measures in isolation, however, are now regarded as inadequate to provide information to help eliminate or reduce the causes of workplace injury or illness because of their numerous disadvantages, such as those listed below.

Good luck vs good management

Unsafe conditions do not always result in accidents or injuries. This can result in a false sense of security and a relaxation of health and safety standards, raising the risk that health or safety will be compromised.

Under- or over-reporting of work-related injuries or disease

Organisations with poor or average OHS performance tend to foster a culture of under-reporting by reacting negatively to reports of high accident and incident rates. This means that, despite appearing to have better safety, these organisations lack accurate information on their safety management, and are therefore limited in the extent to which this information can be used as the basis for improvement.

A preoccupation with outcome performance measures fuels under-reporting. If getting the LTIFR down becomes the focus, rather than improving performance, this can lead to questionable practices which reduce the LTIFR without actually improving OHS management. Both deliberate and unintentional alterations can be made to accident data, sometimes because of pressure to maintain a "good" record. Special difficulties are experienced when the LTIFR is linked to reward systems, in that workers who are paid incentives for long accident-free periods have a tendency to under-report injuries. Claims suppression may occur for a number of understandable, but not always immediately obvious reasons, for example, an unwillingness to be the one who "ruins it for everyone" when an accident ends an accident-free period. This problem tends to be exacerbated in workplaces with many contractors and contractual relationships that convolute the chain of responsibility and reporting.

Sometimes accidents are over-reported, for example in the case of strain injuries sustained outside work, or for which the primary contributing factors were not work-related.

Levels of reporting can change for other reasons as well. For example, there are individual differences — some people report all trivial injuries while at the other extreme some may not even report quite serious ones. Moreover, following a workplace accident, some people may take more time off work than others suffering the "same" injury, thereby biasing severity rates.

There is never a perfect correspondence between injuries and claims of injuries. Depending on the circumstances, people are sometimes more willing and sometimes less willing to claim compensation or injury. Potential claimants may fear negative consequences and can be susceptible to pressure to refrain from claiming for genuine work-related injuries. Alternatively, workers may be unaware of their right to claim, for example, for noise-induced hearing loss which may have been occurring over a long period. A successful claim can sometimes prompt others to follow suit, as a particular type of complaint gains legitimacy amongst the workforce.

For these reasons, rises or falls in the LTIFR can reflect many factors other than improvements or falls in the standard of OHS management.

Indicators of potentially serious incidents are missed

Outcome indicators measure actual injury severity, not necessarily the potential seriousness of the incident. The problem with this can be illustrated by a consideration of two incidents which occurred in a small construction company after a period of a year with no accidents. The first incident involved an employee falling from a ladder and bruising his shoulder and heel. He was not able to work for some weeks and a lost-time injury was recorded. The second incident occurred about a week later, when the company experienced a near miss during a crane lifting operation. After a lifting chain failed, a large prefabricated concrete panel fell two storeys narrowly missing three workers. Though no one was injured, the incident had high potential for injury. Investigation of the near miss revealed that the lifting chains used were damaged and out of service, and they had not been inspected or tested for two years. The company had no inspection and testing procedure in place, and the chains had not been inspected for their currency of service before the lift.

The near miss was potentially far more serious than the lost-time injury, highlighting the inadequacy of outcome indicators and associated measures (such as "number of days since the last lost-time injury occurred") as true indications of effective risk management. Measuring the LTIFR alone provided no information on whether the company had a procedure for inspecting and testing plant and equipment, and whether the procedure was actually carried out.

Delayed outcomes

Outcome indicators do not generally measure delayed outcomes such as long latency occupational diseases which may appear in workers many years after the workplace exposures which caused them. Occupational cancers, such as those caused by asbestos, are in this category.

Trends vs random variation

Most organisations have too few accidents causing injury to be able to distinguish real trends (that is, real improvements or declines in safety performance) from random fluctuations.

Reflections of past practices

Outcome indicators can reflect the outcomes of past OHS practices, if there is a time lag before OHS outcomes reflect changes in OHS practices. For example, purchase of quieter machinery and equipment can be expected to reduce noise-induced hearing loss in the future, but will not be reflected in a drop in numbers of claims in the present.

Feedback for continuous improvement

The purpose of monitoring the organisation's OHS performance is to give management feedback on the basis of which they can direct efforts towards the continual improvement of risk control. Negative outcome indicators such as "number of days since last lost time injury" give management no clue as to which risks may still be present and need to be addressed.

Influence of other factors

In addition to all of the above considerations, indicators such as the number of days lost can also be influenced by factors other than improvements in safety, for example, by improvements in injury management.

For all these reasons, outcome measures alone do not provide a satisfactory indication of the standard of OHS management. Positive performance indicators offer a way of compensating for these deficiencies.

¶7-620 Positive performance indicators

To stimulate improvement in OHS management, indicators are required that measure the processes of safety management. Positive performance indicators (PPIs) focus on how successfully an enterprise is managing and performing in OHS, by monitoring the processes which are intended to produce good OHS outcomes. For this reason, they are sometimes known as process indicators. They are not a substitute for measures of outcome. Rather, PPIs complement outcome measures with additional information in order to give management more useful feedback on the effectiveness of the systems they have in place.

There are advantages in utilising positive indicators of OHS performance to supplement outcome indicators, as PPIs allow for timely and detailed recording of information regarding the process of managing OHS (that is, the type and frequency of health and safety activities that are undertaken by the enterprise). This information can then be used to identify strategies that are working well and other areas in health and safety that require improvement.

Organisations need to develop PPIs that are relevant to their particular requirements — what works for one industry may not work for another. In other words, features of OHS culture and management systems that have the greatest downstream consequences must be identified and indicators developed to show whether these are operating effectively. These indicators should, as a minimum, have the following characteristics:

- they should be accurate, capturing all important elements of performance at the workplace
- they should provide an indication as to how well safety management procedures (systems) are being implemented by both employees and contractors
- they should be results-oriented (but the results must be defined in broader terms than accident and incident frequency rates)
- they must be related to the process of managing safety rather than outcomes alone

- they should be simple so that they can be understood and acted on by everyone in the workplace, and
- multiple measures should be used; no one measure is likely to be sufficient.

There is of course an essential prerequisite to the effective use of PPIs; the performance of an OHS management system cannot be measured until such a system is in place. If the approach to OHS is predominantly reactive and ad hoc, there may not be sufficiently well-defined processes operating for management to gain useful feedback from measuring them.

The process of developing PPIs relevant to the organisation can be considered in a number of stages:

- select the processes to be monitored
- formulate indicators
- document the system and make practical arrangements for data collection, specifying responsibilities, resources, reporting requirements and dates for action and follow-up, and establishing communications as required, and
- review and refine the PPIs.

Selecting the processes to be monitored

The processes intended to produce good OHS outcomes include all aspects of the OHS management system, such as the following:

- commitment and policy
- consultation and planning
- training and competency
- resourcing
- hazard identification
- risk assessment
- risk control
- communication and reporting
- documentation and record keeping
- compliance with new procedures
- accident investigation
- contractor management
- contingency preparedness and response
- monitoring and review of risk management, and
- integration of the occupational health and safety management system (OHSMS) with other management systems.

The focus will vary according to the organisation, its needs and circumstances, but all aspects of the OHS program should be considered for measurement.

Formulating the indicators

The indicators chosen need to be adequately linked to goals, objectives, strategies and targets in order to provide useful information. They should not merely measure "busy-ness". A performance indicator is only as useful as the use that can be made of it, so indicators should be sufficiently sensitive and specific to measure whether the objectives and targets have been achieved, and highlight opportunities for continual improvement. A prerequisite for this is careful formulation of objectives and targets so that performance will be measurable.

PPIs need not necessarily encapsulate all elements of desired performance in a single indicator. They may be designed to cluster around processes, though each one must meet the test of being accurate, simple and dynamic. They are often based on changes in the organisational culture. For example, they may be:

- an increase or a decrease in percentage in a particular factor
- the identification of a level of some sort that the organisation wishes to attain, and
- the establishment of a date by which some achievement is reached.

A suitable positive performance indicator for use in a program to control risks presented by hazardous substances might be, for example "The proportion of hazardous substances for which material safety data sheets have been obtained and risk assessments completed".

These indicators may be qualitative or quantitative, but they should be specific to enterprises and therefore are not necessarily useful for inter-enterprise comparisons.

Documentation

Studies looking at the effectiveness of PPIs have found that inadequate documentation and data can undermine the usability of PPIs. Records need to be available and correctly completed and documentation needs to provide evidence of performance.

Review and refine the PPIs

PPIs need not be constantly changing — there are advantages to using the same indicators over a period of time in order to assess whether improvements in performance are stable and continue to improve. However, it will often be possible to improve on the usefulness of the indicators, both in terms of the selection of processes to be monitored, and in the way the indicator is formulated.

MANAGEMENT REVIEW

¶7-700 Review by senior management

Senior management should regularly review the overall OHS management system to check that it is adequate and suitable. This will require the collection of information, so that management can carry out the evaluation.

The review should be documented and the management system amended if audits or other reports or performance indicators reveal weaknesses in existing arrangements.

Chapter 8

WORKERS COMPENSATION AND INJURY MANAGEMENT

WORKERS COMPENSATION

¶8-100 Introduction

Federal, state and territory legislation has a ''no-fault'' scheme for compensating workers for injuries arising out of their employment, where the employment was a contributing or a significant contributing factor. At the federal level there are two such compensation schemes in place, one applying to federal employees and the other to seafarers.

The rights given to workers by this legislation cannot be limited, excluded or ''contracted out of''.

Each state's and territory's workers compensation Acts are listed in the following table. The contents of the table do not refer to regulations under Acts. Also, the penalties referred to in the table are the major penalties, not the only penalties, for non-compliance. For more detailed information, please refer to your Commonwealth or state or territory workers compensation Acts and regulations.

Coverage	Agency	Legislative instruments	Maximum penalties for non-compliance
ACT	WorkCover ACT	*Workers Compensation Act 1951*	Fines of up to $25,000 for an incorporated body and $5,000 for a non-incorporated body for failing to possess workers compensation insurance.
Federal Government	Comcare Australia Seacare	*Safety, Rehabilitation and Compensation Act 1988*, *Defence Act 1903* (for Defence Force employees from 07/04/94), *Seafarers Rehabilitation and Compensation Act 1992*	Not applicable.
New South Wales	WorkCover Authority NSW	*Workplace Injury Management and Workers Compensation Act 1998*, *Workers Compensation Act 1987*, *Sporting Injuries Insurance Act 1978*, *Workers Compensation (Dust Diseases) Act 1942*, *Workers Compensation (Bush Fire, Emergency and Rescue Services) Act 1987*	Fines of up to 500 penalty units (currently $55,000) and/or six months jail for failing to possess workers compensation insurance. In addition, fines of double the avoided premium, plus other costs. The uninsured employer is also liable for all costs associated with the injury and rehabilitation program. On-the-spot fines and general penalties apply.
Northern Territory	NT WorkSafe, Rehabilitation & Compensation Division	*Work Health Act 1986*	Fines of up to $30,000 for failing to possess workers compensation insurance. General penalties also apply.

Coverage	Agency	Legislative instruments	Maximum penalties for non-compliance
Queensland	WorkCover Queensland	*Workers Compensation and Rehabilitation Act 2003*	Failure to make proper premium payments and periodic returns for a workers compensation policy attracts a fine of 275 penalty units. Continued failure to comply attracts an additional penalty. Claims costs and unpaid premiums may be recovered at 100% of the premium, plus a further penalty of 100% of the premium will be applied. If an employer does not have a current policy and payments have been made to an injured worker, 100% of the payments plus 50% of the total of the payments will be recovered from the employer.
South Australia	WorkCover Corporation of South Australia	*Workers Rehabilitation and Compensation Act 1986*	Failure to pay the levy attracts a fine of up to, but not exceeding, three times the amount of the assessed levy. Penalty interest is payable. Failure to pay a fine on time will attract a further fine of up to three times the original fine.
Tasmania	Workplace Standards Tasmania	*Workers Rehabilitation and Compensation Act 1988*	Failure to have a workers compensation policy attracts a fine of up to 500 penalty units (currently $50,000).
Victoria	Victorian Work-Cover Authority	*Workers Compensation Act 1958*, *Accident Compensation Act 1985*, *Accident Compensation (WorkCover Insurance) Act 1993*, *Accident Compensation (Occupational Health and Safety) Act 1996*	Failure to have workers compensation insurance will result in the employer having to meet the estimated total costs of any claims during the period when the insurance was avoided. A penalty of two times the avoided premium and a fine of an additional 100 penalty units will also be applied.
Western Australia	WorkCover Western Australia	*Workers Compensation and Injury Management Act 1981*	$5,000 for each uninsured worker plus $5,000 for each further week the company is not properly insured (if the offence continues after conviction).

Journey claims

In some cases, an injury which does not arise out of the employment will be covered by workers compensation — this is the category of claim known as a "journey claim". To be covered, the injury must be suffered on a journey between the place of employment and any of the following:

- another workplace
- a place of education
- a place connected with a compensation claim (such as a place of medical treatment or examination), and
- (except in Victoria) the worker's home.

In the Northern Territory, journey claims involving motor vehicle accidents are covered by the no-fault motor accident scheme and are excluded from the workers compensation scheme. Victoria also has a no-fault transport accident scheme which extends to transport accidents on journeys between a workplace and the worker's home.

Defining "injury"

The notion of "injury" is quite broadly defined and can extend to items of personal property, such as clothes or spectacles, which may be damaged as a result of the personal injury. The definition of injury also clearly extends beyond physical injury to encompass psychological injury. However, one by one each government has excluded stress-related injury from workers compensation coverage where it results from "reasonable" management action, such as decisions to, or not to, promote, demote, retrench or discipline a worker.

Specific provisions of the legislation may deal with recognised occupational hazards, such as industrial deafness, brucellosis and asbestos-related conditions.

What does compensation cover?

Workers compensation pays for the injured worker's:

- medical and related expenses (including rehabilitation)
- incidental travel expenses, and
- loss of earnings.

The employer will generally be required to meet these expenses, and will be reimbursed under the workers compensation insurance policy. It should be noted that an employer will not be regarded as admitting liability simply by making these initial payments.

Weekly compensation payments must commence within a short time (up to 14 days, depending on the state or territory) from the time a claim is received, and the worker may be entitled to interest on late payments.

Payments for permanent impairment

If the worker's injury is permanent, an additional amount is generally payable. Traditionally this additional amount was stipulated in a "table of maims", an exhaustive list of particular disabilities with different compensation amounts ascribed to each.

The modern approach to payments for permanent impairment involves the use of "whole person" impairment assessments, where a single amount of compensation is prescribed and the permanently incapacitated worker is compensated according to the percentage of impairment he or she has suffered, assessed by reference to medical guides. This methodology is now used for Commonwealth employees and in New South Wales, Victoria, Western Australia, Queensland and the Northern Territory.

For example, in New South Wales the guides medical practitioners must use to assess permanent injury are New South Wales WorkCover's *WorkCover Guides to the Evaluation of Permanent Impairment* (1st edition, 2001) and the American Medical Association's *Guide to the Evaluation of Permanent Impairment* (5th edition). These guides describe the methods to be used to assess injuries to a range of body systems. This is an attempt to provide greater consistency than was likely before these requirements came into force.

Payments for permanent impairment are based on the whole person impairment figure arrived at using these guides. For example, for the first 1–10% physical impairment, the payment is $1,250 for each one percent. Therefore, an injured person with a 10% whole person impairment receives $12,500.

Assessment using these medical guides is also relevant to common law claims in New South Wales, as access to common law damages is restricted to claims where the degree of permanent impairment is 15% or greater. An impairment threshold of 15% also applies to applications for domestic assistance and for commutation of all benefits.

To be assessed, an impairment should have reached "maximal medical improvement", that is, it should be stable and permanent, or at least unlikely to change by more than 3% in the next 12 months. Impairments are then assessed with a number of different measures, including the injured person's capacity to perform the activities of daily living.

In cases of death, the worker's dependants are entitled to compensation, usually in the form of a lump sum payment and funeral expenses.

¶8-120 Liability and workers compensation

Insurance cover

The liability for work-related injury rests primarily with the employer. Employers (other than approved "self-insurers", including the Commonwealth, state and territory governments) must have a workers compensation policy covering all their workers (including some categories of contractors and volunteers). This policy will generally also cover the employer for any common law liability.

Obtaining workers compensation insurance

In some states and territories, such as New South Wales and Victoria, employers obtain workers compensation insurance by choosing and establishing policies with insurance companies that are licensed or approved by, or act on behalf of, the relevant state or territory workers compensation government agency. There are numerous insurance companies with such approvals. Lists and contact numbers can be obtained from your state or territory workers compensation government agency. In other states and territories, an employer obtains a workers compensation policy directly from the relevant government workers compensation agency. For example, in Queensland employers obtain policies from WorkCover Queensland. Commonwealth departments and authorities deal directly with Comcare Australia which acts as the insurer.

Certain employers may choose to become self-insurers and accept all of the financial costs and liabilities for their own workers compensation and rehabilitation programs. However, this status is normally only available to large employers which satisfy certain requirements. These requirements are usually based on economic criteria, plus additional criteria, such as the number of workers an organisation has in a particular state or territory. The criteria vary between each of the states and territories. Commonwealth government departments and authorities are not permitted to self-insure.

To become a self-insurer employers need to obtain relevant application forms and assessment details from their state or territory government workers compensation agency.

Premiums and policy pricing

The price employers pay for workers compensation policies in Victoria, New South Wales, South Australia, Western Australia and Queensland is usually based on:

- the size of each employer's payroll (which includes benefits such as superannuation, use of motor vehicles and other fringe benefits)
- the duration for which the policy will apply (usually 12 months)
- the claims history or experience of the employer — the time period required for an employer to establish a history or experience is different in each of the above-mentioned states, and
- a premium (levy) rate or order — the rate or order is different in each of the above-mentioned states.

The premium (levy) rate or order allocated to an organisation is a percentage of the organisation's payroll. An industry classification system is used to arrive at an industry risk rating which is then used to determine the premium rate (levy) or order. The risk rating is based on all claims in a given industry sector for a set period of time. The duration of this set period of time differs between each of the states mentioned above.

The result is that high-risk industries have higher premium (levy) rates and therefore pay higher premiums (levies) than lower-risk industries.

For example, a high-risk industry such as the housing construction industry might have a premium rate several times higher than the average rate for that state. Similarly, a low-risk industry such as private secondary schools might have a rate only half the average (all-industries) premium (levy) rate for that state.

Organisations operating branches in Victoria, New South Wales, South Australia, Western Australia and Queensland need to understand that premium (levy) rates vary between these jurisdictions. For example, the levy rate for nursing homes in one state may be 3.2% of the payroll, while in another it could be 7.6%.

When obtaining workers compensation insurance, the insurance company or government agency allocates the industry classification (which is used to determine the industry risk rating and therefore the premium (levy) rate). If the employer disagrees with the allocated industry classification and the resultant premium (levy) rate, the employer may appeal to the relevant statutory authority for a review. This should be done as soon as practicable after the policy is issued because a time limit for appeals may apply (for example, 21 days in New South Wales).

The procedure for determining industry classifications varies between Western Australia, Victoria, South Australia, New South Wales and Queensland. Western Australia uses the *Australian and New Zealand Standard Industry Code* (ANZSIC), Victoria and South Australia base their classifications on the *Australian Bureau of Statistics Australian Standard Industry Code* (ASIC) and NSW and Queensland have systems of their own.

In some states and territories certain industries are excluded from the central workers compensation scheme, such as the coal industry in New South Wales.

Apart from Western Australia, in states and territories where there is full private underwriting, such as in Tasmania, the Northern Territory and the Australian Capital Territory, state and territory allocated industry risk rates are not used to calculate premiums (levies). In these states and territories market forces play a major role in determining premiums (levies).

For Commonwealth departments and authorities, Comcare Australia determines premiums (levies) based on a prescribed amount modified by a penalty/bonus system which takes into account each agency's claims performance.

In New South Wales, there is a premium discount scheme under which participants can earn discounts on their workers compensation insurance premium (up to 15% in the first year), for a period of three years, by making their workplaces safer and improving their injury management systems. WorkCover NSW reports that the scheme has delivered millions of dollars worth of premium rebates to thousands of employers in the state.

A penalty/bonus system also operates in some states and territories where variations of up to 50% of the levy (premium) are available to employers who meet certain criteria.

When employers provide estimates of their claims experience, cost and frequency of accidents, and payroll to the insurer or government agency, it is essential that the estimates are accurate. Employers are required to complete forms or statutory declarations in relation to payroll, claims experience and accident information. This information is legally binding, and if there is any significant variation between the statements made by the employer, and what actually occurs, it may be used against the employer in legal proceedings. Underestimating these figures will lower the employer's premium (levy) in the short term, but fines and penalties will be applied to employers found engaging in such practices (for example, fines of up to $11,000 can be applied in New South Wales).

Evading proper payment of premiums (levies) means that some employers end up carrying costs that would otherwise be distributed to all employers. The most appropriate method for employers to reduce their workers compensation cost is to improve their occupational health and safety and injury management performance.

An industry association, union or similar organisation can instigate a cooperative, industry-wide program to improve OHS and rehabilitation performance. In time, risk management strategies leading to injury prevention, and rehabilitation leading to a more rapid return to work of injured workers, will decrease industry workers compensation costs. This, in turn, will produce a decrease in the industry risk rating and therefore a decrease in the premium (levy) rate.

Individual employers can implement risk management programs within their individual workplaces that will improve OHS performance and injury management. Over time, employers will see a decrease in workers compensation costs because there will be a decrease in claims experience and therefore a reduced history of claims. Employers will then be in a position to negotiate with the insurer or government agency to decrease the cost of their workers compensation policy.

Ultimately, all improvements in workplace safety should produce financial returns to the employer.

Lodging a claim

In the Commonwealth and all states and territories, an employee is eligible to make a workers compensation claim if he/she sustains an injury (physical or mental) that arises out of the course of his/her work.

The procedure for making a claim differs slightly between the Commonwealth, the states and the territories, but generally all procedures aim to ensure that:

- a worker notifies the employer as soon as possible after suffering an injury — this usually requires the completion of an injury register or accident form, but under certain conditions notification may also be verbal (for example in Victoria and Tasmania)

- the injured worker sees an accredited medical practitioner
- the accredited medical practitioner completes a workers compensation medical certificate — usually in a format approved by the government workers compensation authority
- in the case of serious injury or accident, the employer notifies the local OHS agency and insurer as soon as is practicable (usually within 48 hours) — a serious injury normally constitutes but is not limited to
  - the need for medical treatment within 48 hours of being exposed to a substance
  - the need for immediate hospital treatment as an in-patient
  - amputation
  - serious head injury
  - serious eye injury
  - separation of skin from underlying tissue (for example, scalping or de-gloving)
  - electric shock
  - spinal injury
  - serious laceration
- the insurer is notified of a minor claim, even though the employer is usually fully liable for the associated costs
- the worker completes a claim form (usually a prescribed format) and submits it with the medical certificate and any supporting documentation to the employer — the employee should retain copies of all documents submitted
- the employer accepts and then acknowledges the worker's claim in writing
- the employer sends the claim form, medical certificate and associated documentation to the insurer, which assesses the claim
- the insurer notifies the employer as to whether liability for the claim is accepted — if liability is accepted, the employer is required to make payments such as wages and medical costs for the amount and/or period of the prescribed excess
- the insurer continues payment from the "excess" threshold, and may be liable to pay compensation for specific permanent injuries or disease
- the insurer at its own expense, requires the employee to attend medical examinations
- the insurer advises a worker within a specified time limit such as a maximum of 42 days in New South Wales, or 14 days in South Australia if a claim is to be denied

- the employee has the right to approach a conciliation tribunal to appeal a claim denial — if the matter is not resolved at conciliation, it may then continue to the workers compensation court and/or litigation under common law (except in Victoria, the Commonwealth, South Australia and the Northern Territory, where common law rights have been extinguished under certain circumstances), and
- once the insurer has accepted the claim, the worker receives appropriate financial compensation, and when the worker is fit to do so he/she is entered into a suitable rehabilitation program which includes a graded return-to-work plan.

Information and forms relevant to your Commonwealth, state or territory's workers compensation system should be sought from either your insurer, or state or territory government workers compensation agency.

Where the employer is uninsured

If an employer is uninsured, workers compensation will be paid by a statutory fund through a nominal insurer (much along the lines of the nominal defendant scheme for motor vehicle accidents where the driver at fault is unidentified or uninsured). The fund may then sue the uninsured employer for the amount paid out. Where private insurers are involved, a statutory fund exists for the contingency that the employer's insurer has ceased to exist or is otherwise unable to provide an indemnity.

"No-fault" liability

The "no-fault" system means compensation will be paid regardless of whether or not the employer was negligent or breached some other law. Similarly, any fault on the part of the injured worker will generally be disregarded. However, where the work-related injury has occurred as a result of the worker's serious and wilful misconduct, compensation will be denied unless the injury results in death or serious disability.

In some workers compensation legislation, compensation in cases of intoxication and deliberately self-inflicted injury, which may amount to serious and wilful misconduct in any event, is expressly excluded.

Common law rights

Workers compensation and common law damages are quite different things. Some legislation has completely abolished a worker's right to sue his or her employer for common law damages. Other legislation has limited common law rights by either or both of the following means:

- restricting common law access by insisting that the injury must exceed an arbitrary level of seriousness before a case can be brought, and
- restricting common law recovery by limiting the amount of damages that can be awarded.

Where workers recover common law damages, they are required to repay, up to the amount of the damages received, any workers compensation that was received in respect of the same injury.

Other liability

As previously noted, workers compensation is confined to personal injury and death. Therefore, any abolition or restriction of the worker's common law rights to sue is generally confined to personal injury and death. As a result, it is advisable for employers to insure against other types of liability, such as property damage or financial loss not resulting from personal injury. Even if such an action ultimately fails, the costs of defending it can be high.

Third party also responsible for injury

A worker's injury may arise out of the employment but may also have been caused by someone other than the employer. For example, a worker who is required to work away from the employer's premises might be injured by the dangerous condition of the premises where they are sent to work, or by the negligent driving of another motorist en route to the premises. Defective machinery in the workplace is also a source of such third-party liability, as there may be a claim against the manufacturer.

Any abolition or restriction of workers' common law rights against their employers will generally not extend to such third parties who are at fault. If so, the employer or workers compensation insurer will be entitled to recover from the third party the compensation paid to the worker, or the proportion of it which reflects the extent of the third party's contribution to the injury. If not, the worker may be able to sue the third party but if successful will be required to repay some or all of the workers compensation received.

¶8-130 Duties under workers compensation legislation

The various workers compensation schemes cast legal obligations on the following parties:

- workers
- employers
- the body which administers the workers compensation scheme
- insurers (if the workers compensation body is not also the insurer), and
- self-insurers.

Many of these obligations revolve around disclosure and the giving of relevant information — including exchange of information between states and territories.

Employees' duties

Workers are obliged to promptly report injuries to their employers. Throughout Australia, workers are required to make a claim for workers compensation within six months of the injury. This period may be extended in certain cases — for example, in the case of an occupational disease (such as mesothelioma) which has a long latency period.

Disclosure of information

On commencing employment, workers are generally required to disclose existing disabilities and illnesses, or at least not to deliberately withhold this information. Failure to disclose relevant information may result in their losing the right to compensation in relation to aggravation of undisclosed pre-existing conditions.

Workers in receipt of weekly compensation payments have a duty to notify their employers of any change in circumstances.

Workers must act honestly in making a claim for compensation. A claim for compensation will usually need to be accompanied by a medical certificate. Where a worker provides false or misleading information in support of a compensation claim, or makes a fraudulent claim, the penalties are heavy and may include imprisonment.

Employers' duties

Once they receive a claim for workers compensation, employers must promptly notify their insurer. Employers cannot refuse to receive a claim for workers compensation.

Information and records

Employers are generally required to disclose the name of their workers compensation insurer on request, and to provide workers with information relating to their rights and the making of claims. In some parts of Australia employers are required to display a summary of the workers compensation legislation in a conspicuous place.

The workers compensation legislation also requires employers to keep certain records relating to their employment of workers, such as wage records and perhaps a register of injuries.

When a new employee commences work, it is advisable for employers to ask them about any pre-existing disabilities, illnesses or conditions which may impact on potential workers compensation claims.

Paying benefits to employees

The employer is liable to provide a direct payment to an injured worker, during the initial stages of the worker's injury. This direct payment is referred to as the excess. The degree of the liability varies between the states, territories and the Commonwealth. The following table outlines the employer liability in regard to the payment of the excess, the injured worker's payment rates (wage rates and entitlements) and the payments the insurer or government workers compensation agency continues to pay after the excess has expired.

Note: < = means less than or equal to; NWE means normal weekly earnings; AWE means average weekly earnings.

Jurisdiction	Employer excess liability	Payment rates
ACT	Nil. The scheme covers all costs from the first day of incapacity.	< = 26 weeks — sick pay/NWE. Where conditions of employment include sick leave entitlements, payments are equivalent to sick leave rates. Where conditions of employment do not include sick leave entitlements, payments are equivalent to NWE which are equivalent to regular pay rates. No overtime or additional allowances are included.
Commonwealth and Seacare	Nil. The scheme covers all costs from the first day of incapacity.	< = 45 weeks — NWE up to a maximum of $1,140.15. NWE includes an average of the overtime worked on a regular basis, but excludes expenditure reimbursement or any special allowances. Income from external employment is also included in NWE, if the injured worker was employed in part-time Commonwealth employment at the time of the injury.
New South Wales	Category A employers (annual premiums > $10,000) pay the first $500 of weekly payments for each claim. Category B employers (annual premiums < $10,000) pay the first $500 of payment or payment of excess surcharge on premiums of 3%*. **correct as at 1 October 2005. Changes to the scheme are expected after 31 December 2005. Check with WorkCover NSW for details.*	< = 26 weeks — basic award rate or enterprise agreement rate, or the worker is paid 80% of the AWE, with a maximum weekly payment of $1,420.50*. AWE is estimated by assessing the last 12 months of wages, including amounts payable under common industry practice, but excluding shift allowances, penalty rates and overtime.
Northern Territory	The day of the injury.	< = 26 weeks — NWE. NWE includes salary; overtime; shift penalties; climate, district, leading hand and qualification allowances; and any regular non-cash entitlements (for example, loss of the use of a motor vehicle).

Jurisdiction	Employer excess liability	Payment rates
Queensland	The day of the injury paid by the employer, and the employer pays a maximum of $500 as a lump sum. If the weekly wage is less than $500, the employer pays NWE less $1. There is an option for the employer to buy out the excess at 5% of the premium or $10, whichever is larger.	< = 26 weeks — 85% of NWE for workers under an industrial agreement. Workers not covered by an industrial agreement are paid 70% of Queensland Ordinary Time Earnings (QOTE), or 85% of NWE (whichever is greater). NWE includes salary, overtime and any regular non-cash entitlements (for example, loss of the use of a motor vehicle). It does not include one-off periods of overtime.
South Australia	The first two weeks of incapacity per worker per calendar year.	< = 52 weeks — AWE to a maximum of two times the state average. AWE includes salary, overtime and any non-cash entitlements.
Tasmania	The first five working days of each injury and the first $200 of other benefits/ costs.	< = six weeks — NWE for the previous 12 months, or the rate of pay at the time of accident. NWE includes salary and any regular non-cash entitlements (for example, loss of the use of a motor vehicle). Overtime is not included.
Victoria	The first 10 days of incapacity, and reasonable medical costs to a prescribed amount.	Injuries sustained post-12 November 1997 — 95% of the pre-injury average weekly base pay to $1,150/week is paid for the first three months. For permanent impairments, payments are determined through a process of "whole person impairment" assessment.
Western Australia	Nil. The scheme covers employers from the first day of incapacity.	< = four weeks — pre-injury rate of pay to a maximum of state AWE, or award wages plus any regular over-award payments excluding overtime (whichever is greater).

Rehabilitation obligations

A major characteristic of modern workers compensation schemes is the emphasis they now place on the rehabilitation of injured workers. In some parts of Australia, compensation will extend to modifying the workers' homes or motor vehicles to accommodate their disability.

Employers (other than small employers in some parts of Australia) are required by the legislation to have rehabilitation policies in place. For further details, see ¶8-500 in this chapter.

Enforcement activities

The statutory bodies which administer the workers compensation schemes employ inspectors to police the requirements of the legislation. These inspectors have certain powers to demand and inspect various documents. In some jurisdictions the inspectors are empowered to issue on-the-spot fines for breaches of the legislation.

Dispute resolution

If a workers compensation claim is disputed, most legislation insists that some form of conciliation take place before legal proceedings can be commenced. Even in circumstances where this does not occur as a matter of course, the court or tribunal determining the claim may refer the dispute for conciliation.

Each of the workers compensation schemes also enables medical questions to be referred, and most of the schemes have established medical panels for this purpose.

No good faith duty

Unlike American law, Australian law does not recognise any duty of "good faith" owed by an employer or an insurer to a worker making a claim for compensation.

The relationship between the insurer and the employer is contractual and the worker is not a party to this contract. The employer's relationship with the worker is a much broader one, and this may give rise to further duties. For example, most workers compensation schemes do not themselves prohibit the sacking of a worker who has made a claim for compensation, but such a dismissal may be unfair or unlawful in accordance with other industrial laws or awards.

Where the insurer is a statutory or government body, it will be subject to the principles of administrative law. These principles provide the person who may be affected by a decision with the right to be heard. They also require the person making the decision to give reasons for the determination. Therefore, even where there is no right to appeal from a decision, it may be overturned if the employee has not been given a fair hearing.

INJURY MANAGEMENT

¶8-500 What is injury management?

While the work of the OHS manager is primarily focused on preventing accidents and ill-health in the workplace, this role also extends to dealing with situations where employees are injured or suffer illness through their work. The majority of employees who are ill or injured simply return to work after basic first aid and medical treatment, but a minority who are more seriously affected should have their return to work managed through a process of occupational rehabilitation.

The need for injury management arises where ill or injured workers' ability to carry on their duties has been impaired. The primary aim of injury management and occupational rehabilitation is to return workers to their full workplace duties as soon as possible after an illness or injury. To ensure that this can be done effectively immediately following an employee's untimely absence, workplaces should develop their own rehabilitation policy and program. These:

- outline the organisation's commitment to the injury management/ rehabilitation process
- designate the necessary resources for the process, and
- set out the procedures to be observed following a workplace injury or illness.

The process of rehabilitation must be undertaken safely, that is, without risk of causing further injury or illness to the worker. For this reason, an individual return-to-work plan must be developed, tailored to the worker's needs and capabilities following the injury or illness and also to the capacity of the workplace. This may mean, for example, that the worker returns to lighter duties initially, or that his/her hours of work are shorter than the normal work shift or routine.

Injury management and occupational rehabilitation

In some states and territories "workplace injury management" is the term most commonly used in relation to the return to work of an injured worker. It has a broader meaning than "occupational rehabilitation", encompassing treatment, claims management and other employment practices as well as rehabilitation. In this section, principles and procedures are described using the terms "injury management", "rehabilitation" and "return-to-work". It should be noted, however, that there are some inconsistencies across the different states and territories in the definitions of some terms, so readers should check information provided by their own jurisdiction for local terminology and meanings.

Benefits of injury management

There are clear benefits for both the employer and the employee from successful rehabilitation.

The main beneficiary from good injury management practices is the injured worker. Long-term loss of the ability to work usually leads to a loss of status and morale, and sometimes financial difficulties. The best way to address these problems is to engineer a return to work as soon as it is safely possible.

There are also significant benefits for employers. Effective injury management will mean that employees are able to return to work sooner, with the consequential savings in workers compensation costs. As it is the minority of long-term injuries that make up the large majority of workers compensation costs, cases of organisations halving their workers compensation costs have been reported. In addition, there are intangible benefits, such as increased staff morale.

Legislative encouragement of rehabilitation

In recent years the workers compensation legislation in most Australian states has been amended to place increased emphasis on rehabilitation. In most states employers are now required to develop rehabilitation programs and/or to provide suitable duties for injured employees.

Steps in injury management

To gain maximum benefit, the rehabilitation process should be activated as soon as the employer is notified of the worker's injury or illness. The basic steps in implementing an individual return-to-work plan include:

- assessing the worker's needs and abilities, which involves liaising with any treating medical practitioner(s) and the worker
- developing a return-to-work (RTW) plan, which identifies the timing of the worker's return together with specific responses to the worker's needs, such as identification of suitable duties and work periods
- gaining agreement on the RTW plan from all parties involved (the worker, workplace management, work supervisors, any relevant union representatives and treating medical practitioners)
- implementing the RTW plan and monitoring progress, and
- reviewing the effectiveness of the RTW plan and making any necessary adjustments.

The rehabilitation process ends when the worker has returned to his/her full pre-injury or illness duties, or those duties which are agreed as a satisfactory long-term outcome. Where it is not possible to rehabilitate the employee back to former duties, an attempt must be made to find a suitable alternative position for the worker within the same workplace. If this cannot be achieved, a further process of work skills assessment, counselling and career redirection should be undertaken.

¶8-520 Who manages the rehabilitation function?

Injury management responsibilities are usually undertaken by a nominated in-house "rehabilitation coordinator" (or "return-to-work coordinator"). This person may also be the OHS manager or the human resources manager within the workplace.

The role of occupational rehabilitation coordinators is to:

- assist their employers in meeting their legal obligations with regard to rehabilitation
- manage the workplace rehabilitation policy/program and individual RTW plans
- liaise between workers, their work supervisors, workplace management, any relevant union representation, treating medical practitioners,

specialist rehabilitation providers (where these are needed) and the workers compensation insurer, and

- keep records as required under the injury management/rehabilitation guidelines or requirements of the relevant authority (that is, the agency listed in the table at ¶8-100).

Following up an injury or illness

Once the workplace injury management policy and program are in place, the usual process which follows the employer's notification of a worker's absence due to injury or illness is for the rehabilitation coordinator to:

- make contact with the employee involved, offer any available assistance and make an initial determination on whether rehabilitation is necessary or possible
- arrange for a professional assessment of the employee's needs and capabilities
- develop a written RTW plan for the worker which describes return-to-work arrangements
- monitor the employee's recuperation and oversee the employee's return to suitable duties at the workplace as soon as possible
- monitor progress of the return to work and supervise/adjust the arrangements to ensure the employee returns safely to pre-injury or illness duties (where this is possible), and
- record the entire process as required.

Participation of the injured worker in decision making

Too often injured workers are made to feel that they have no control over decisions concerning their health. An injured worker who is part of the decision-making process is more likely to be committed to a successful outcome.

Injured employees must retain the right to choose whether to participate in rehabilitation treatment, and the right to choose their doctor and rehabilitation provider. The legislation in some states, however, provides disincentives (a reduction in benefits) where people refuse to participate in a rehabilitation scheme.

¶8-530 Suitable duties

Suitable duties are the cornerstone of any workplace injury management program. They allow the employee to return to the workplace, increasing fitness and reducing any depression or "compensation syndrome". For the employer, suitable duties can greatly reduce the working days lost due to injury.

Suitable duties are tasks designed specifically for an injured worker in order to allow an early return to work. The work must be carefully matched to an employee's capabilities so that it facilitates recovery while not risking further injury.

The duties provided may be some of the employee's previous tasks, or they may be the previous tasks performed at a slower speed or for fewer hours per day, or they may be completely different tasks. Any planning of suitable duties should be done in consultation with the injured worker, his/her doctor and the employees and supervisors who will be directly affected.

It is important to monitor and upgrade a program of suitable duties, to ensure the worker is progressively moving towards full recovery. Gradually increasing the time spent at work is one common way of achieving this.

Problems can arise when suitable duties are provided in an area other than the injured worker's usual area of work. The injured worker may be working in an area covered by another union, and may be on a different rate of pay from other employees performing the same work. These issues should be addressed in the organisation's rehabilitation policy. It is important to ensure that a rehabilitation program does not create any risks/hazards for other employees by making changes to their jobs to accommodate an injured employee.

In some states there is a legal requirement to provide suitable duties — see ¶8-570.

¶8-540 Using external rehabilitation providers

In certain cases, due to the complexity or duration of the injury or illness, or the treatment needed, it may be necessary for the rehabilitation coordinator to call for expert help — via a rehabilitation provider. External providers may also be necessary or desirable in other circumstances — such as where the type of work involved is likely to exacerbate the injury. An example here might be a worker suffering from occupational overuse syndrome (OOS, formerly known as repetitive strain injury or RSI) whose job involves undertaking repetitive tasks. In such cases, the rehabilitation provider will be able to assist in modifying the task, the work environment or the way in which the work is performed.

Rehabilitation providers offer a range of specialist services including medical/health assessment, treatment and management (which may include both the physical and psychological aspects of the injury or illness), ergonomics and workplace or task modification consultancy, vocational assessment and career redirection (where necessary), and other advisory services.

The need for a rehabilitation provider is usually determined by the treating doctor — although, in some cases, the rehabilitation coordinator may recognise the need first. The workplace should have a list of providers available as part of its rehabilitation policy/program. In fact, employers which meet certain "size" criteria stipulated in the relevant legislation are required to nominate in their policy those occupational rehabilitation providers whose services will be used when the need arises.

The choice of the correct provider in each individual case will depend on factors including the range of specialist services required and the proximity of the provider's premises/services to the workplace and/or the ill or injured worker's home.

Role of the provider

Once appointed, the provider will be responsible for:

- contacting and maintaining ongoing communication with the treating doctor
- assessing the employee's rehabilitation requirements
- arranging for, and monitoring, any therapy and/or counselling needed
- developing a written rehabilitation plan for the employee which includes the nature and duration of any treatment to be undertaken and the RTW plan (outlining specific details such as the hours of work, the range of tasks which can be undertaken and the timing for undertaking increased duties until the return to full duties is achieved)
- monitoring the rehabilitation plan, ensuring that it is followed and, where necessary, adjusting the plan to take account of changes in circumstances (such as faster or slower than anticipated recuperation of the employee), and
- liaising between the in-house rehabilitation coordinator (or other employer representative), the worker, the treating doctor, any career or other counsellors and the workers compensation insurer/manager.

Professional accreditation

Rehabilitation providers are subject to strict professional standards and must be approved or accredited by the government authority overseeing the rehabilitation arrangements at the federal, state or territory level. A list of approved providers will be available from your region's workers compensation authority.

Connection with workers compensation

The injury management function is closely linked to workers compensation. When a worker lodges a claim for compensation in relation to workplace injury or illness, this claim is processed either by the employer, the employer's workers compensation insurer or a government authority. Where the claim is accepted, the insurer will cover the costs of the worker's salary during time off work, medical and related expenses and also the cost of rehabilitating the worker.

For this reason, the workers compensation claims management and the occupational rehabilitation program must link up — through the insurer or claims manager liaising closely with the rehabilitation coordinator or provider.

The claims manager must be aware of the purpose and progress of RTW endeavours. Likewise, the rehabilitation coordinator needs to understand the compensation issues which directly impact on the employee's salary arrangements and RTW incentives. For example, the employee's entitlement to compensation benefits may be jeopardised if the employee unreasonably refuses to cooperate in a rehabilitation program. Therefore, there needs to be ongoing personal contact and exchange of written progress reports between the parties responsible for the two processes.

¶8-550 Employers' key duties

Duties relating to rehabilitation/injury management are usually included in the federal, state or territory workers compensation legislation. Although specific details may vary, the principal requirements are similar throughout Australia.

Most workers compensation legislation requires certain employers (defined by size) to establish and maintain a rehabilitation policy/program to ensure that the process of assisting injured employees takes place.

Which employers must rehabilitate?

The criteria defining which employers must establish a rehabilitation/return-to-work program vary. Examples of the thresholds at which rehabilitation programs must be developed include:

- in Victoria, employers with rateable remuneration exceeding $1 million qualify
- in New South Wales and Tasmania, employers with 20 or more employees must comply, and
- in South Australia, employers with 10 or more employees qualify.

Minimum requirements

The minimum components of a rehabilitation program include:

- preparing a statement of the employer's rehabilitation and RTW policy
- providing an outline of the specific steps or procedures which will be followed if an injury occurs in the workplace
- appointing a rehabilitation coordinator who will ensure that the steps outlined take place, and
- determining and allocating resources (including nominating rehabilitation service providers) which will be used to assist with the rehabilitation of an injured employee if the need arises.

It is advisable for the rehabilitation program to be integrated into the organisation's overall OHS risk management program, and for the agreed procedures to be documented in reasonable detail so that rehabilitation efforts are cost-effectively and consistently applied for all employees.

Rehabilitation costs

Rehabilitation expenses are special costs which the employer/insurer is liable to meet — in addition to other compensation costs. The costs are either specifically mentioned as part of the worker's entitlement to compensation or are included in the definition of "medical services" or similar for which the employer or insurer is liable.

Permanently incapacitated workers

The major aim of rehabilitation services is to maintain injured workers at work or, alternatively, to secure the workers' return to work (ideally to his/her original workplace and job) as soon as possible. This may be difficult where workers remain partially incapacitated long-term and can no longer perform their former duties.

Under some legislation the worker is protected from dismissal following injury for a specified period and/or from losing his or her entitlement on return to work if the return is only for a short period and is unsuccessful. There is also provision in some legislation for subsidies to be provided for work trial and work placement programs for partially incapacitated workers.

Workers' duties

In general, employers have a responsibility to provide rehabilitation (or support), and employees are obliged, where medical and other evidence indicates it is desirable, to cooperate:

- in rehabilitation and training programs, and
- generally, for example by undergoing periodic medical examinations to ascertain the current nature and extent of the injury and capacity for work.

¶8-560 Coordinating the rehabilitation function

Developing the rehabilitation policy

The first step in managing the rehabilitation function is to develop a workplace injury management/rehabilitation policy. This is a statement of commitment by both the organisation and its employees to the rehabilitation of employees following injury or illness. A sample policy is included later in this section.

Policy commitments

At a minimum, the policy should include the following commitments:

- The company will offer immediate medical treatment if an injury occurs.
- Rehabilitation services will be available as soon as is necessary to achieve a prompt return to work.
- Return to suitable duties will be facilitated as soon as medical advice suggests it is appropriate.
- Appropriate assistance, including modification of jobs and the workplace, will be offered to allow the employee to return to suitable work.
- Suitable employment opportunities will be offered, wherever possible, if an employee is unable to return to pre-injury duties.
- Where the period of absence from work requires it, an RTW plan will be developed with the involvement of the injured employee and the employer representative.
- The employee's RTW plan will be regularly monitored, reviewed and adjusted to reflect the changing needs of the employee and the workplace.

Managing the policy

The policy should be:

- developed in consultation with the employees
- endorsed by a senior executive of the organisation
- broadly communicated throughout the organisation
- accessible to all employees
- kept up to date, and
- integrated with injury prevention and claims management as part of the company's overall risk management policy.

Implementing the occupational injury management/rehabilitation program

In some legislation, "rehabilitation program" is used to mean the activities planned for the individual employee's return to work. In others, "rehabilitation program" means an employer's overall, organisation-wide strategy to implement individual RTW plans — this is the meaning of the term applied here.

Role of the program

The rehabilitation program is a documented plan of action outlining the procedures that will be followed when any employee is injured at work in order that:

- the injury is quickly and effectively treated
- the employee is returned to "suitable duties" in his/her original workplace as soon as possible, and
- the other objectives of the rehabilitation policy are met.

Fleshing out the program

The program must be general in scope so that it can be applied to all situations where ill or injured workers may need rehabilitation. It should set out clearly the steps to be taken from the time when the employer is informed of the employee's injury or illness right through to the final review which determines that the employee has been restored as completely as possible to the employment situation which existed prior to the injury or illness.

As the rehabilitation or RTW coordinator is a key person in the injury management process, this person's name must appear in the documentation of the rehabilitation program. The program should also outline the external rehabilitation providers which will be used if and when the need arises. These providers should be accredited by the relevant workers compensation authority.

Other issues which should be addressed by the program include:

- the means of consultation on injury management issues which is to be used between the employer and employees (including any union representation), and
- the mechanism to be used in dealing with disputes concerning injury management issues.

A copy or summary of the rehabilitation program should be displayed prominently in the workplace.

Developing rehabilitation/return-to-work plans

Most legislation requires employers to develop a rehabilitation/return-to-work plan for injured workers where the duration of an employee's absence from work with a work-related injury exceeds a specified period (such as 20 days). Implementation of the plan must usually be commenced within a certain time frame (for example, within 10 days of reaching the 20-day absence point).

Details of individual plans

The RTW plan is a document listing all the activities agreed to by the employee involved and by the employer's representative who approves the activities. It includes:

- the name and contact details of the employee, workplace supervisor, treating doctors, rehabilitation service providers and the RTW coordinator
- the relevant details of the injury and medical restrictions
- a list of activities to be undertaken during the plan's operation
- an offer of suitable employment at the workplace and an estimated RTW date (if applicable — some injuries do not involve an absence from work)
- actions to be taken to prevent the injury from recurring (such as modification of the employee's workstation)
- the date when the plan will be reviewed, and
- the signatures of the employee and the employer representatives signifying approval of the plan's implementation.

Providing suitable duties

In general, the employer has a duty to:

- re-employ workers who have had a compensable injury by providing them with suitable duties, and
- keep workers' jobs open for a reasonable period following the injury or illness (from six to 12 months).

However, if, due to the nature or extent of the injury, the worker cannot be returned to the original position, some assistance should be offered to obtain a suitable job.

The requirements in each state are mentioned at ¶8-570.

Further references

Guidance Note for Best Practice Rehabilitation Management of Occupational Injuries and Disease, National Occupational Health and Safety Commission, 1995.

Uniform Guidelines for the Accreditation of Rehabilitation Providers, National Occupational Health and Safety Commission, 1995.

Sample policy

The sample injury management/rehabilitation policy below may be used as a basis for a policy tailored to the needs of a particular organisation.

Injury management/rehabilitation policy

The company is committed to preventing illness and injuries at the workplace by providing a safe and healthy working environment for all employees. It is recognised that injury or illness may still occur and therefore all incidents will be reviewed and steps will be taken to prevent recurrence.

The company believes that occupational rehabilitation is of benefit to everyone and should commence as soon as possible following injury or illness. Furthermore, no person being rehabilitated will suffer financial loss or prejudice in any way.

Early reporting of injury and illness is encouraged.

The company will ensure access to good first aid and high quality medical care. Accurate medical diagnosis and assessment will be followed by early intervention from a rehabilitation service provider if necessary.

Every effort will be made to assist people in an early, safe return to meaningful and productive work in consultation with their treating practitioners.

Suitable duties will be provided by the company; where this is not possible, early referral to a rehabilitation unit will be facilitated.

A graduated return-to-work program consistent with medical advice will be followed. Each person will be given a written return-to-work program. The rehabilitation coordinator will assist in this process by providing the necessary link between treating practitioners, rehabilitation service providers and the workplace.

The company has consulted with employee and union representatives; there will be no inter-union disputes (demarcation disputes) arising because some people may have to be rehabilitated with alternative duties outside their usual job classification. However, adequate training for such alternative duties will be given to ensure that safe working practices are followed.

The company's rehabilitation coordinator is

....

The company's chosen rehabilitation provider(s) is/are:

....

(name and address)

....

The injured employee retains the right to choose his/her own treating doctor and to choose an alternative rehabilitation provider.

All employees on site have an important role to play to ensure the best possible outcome for their injured colleagues; successful injury management requires everyone's involvement and commitment.

Employee Representative:

....

Chief Executive Officer:

....

¶8-570 Rehabilitation systems — state by state

The following provides a brief description of the government regulation of rehabilitation in each state. Further information can be provided by the relevant authority in each state.

New South Wales

The *Workplace Injury Management and Workers Compensation Act 1998* introduced the concept of injury management — a "coordinated program of treatment, claims management, rehabilitation and employment practices to achieve a safe, timely and durable return to work for injured workers" — as a cornerstone in the management of workplace injuries in New South Wales.

The key features of the system require insurers to develop injury management programs governing their procedures for managing workers compensation cases, and employers to develop RTW programs that comply with their insurers' injury management programs. Employers' RTW programs must address the commitments and procedures set out in the WorkCover NSW *Guidelines for Employers' Return to Work Programs*, that apply in total to larger employers, ie those with a basic tariff premium of more than $50,000, or self-insurers, or employers with more than 20 employees, who are insured with a specialised insurer. All other employers (who do not meet any of the above criteria) may use a WorkCover NSW standard workplace RTW program.

The system also requires employers to notify their insurer of significant injuries (those requiring more than seven days away from normal duties) within 48 hours. Workers with significant injuries must nominate a doctor to oversee their injury management. The insurer, in consultation with the employer and the nominated treating doctor, will develop an injury management plan for individual workers with significant injuries. Employers are expected to make suitable duties available to injured workers.

Victoria

Rehabilitation in Victoria is administered by the Victorian WorkCover Authority.

Certain employers (those with a rateable remuneration of over $1 million annually) are required to establish and maintain an occupational rehabilitation program. If a worker is incapacitated for 20 or more days, the employer must prepare an RTW plan for that worker, and nominate an RTW coordinator.

Employers must keep jobs open for a period of 12 months to provide suitable employment for injured workers. Workers may have their weekly benefits stopped if they refuse to make reasonable efforts to return to work, refuse to accept a suitable job offer, refuse to undergo training or refuse an assessment as to rehabilitation prospects.

Queensland

Queensland employers with 30 or more workers at a workplace must have a workplace rehabilitation policy and procedures in place, review the policy at least every three years (and provide WorkCover Queensland with written evidence, in the approved form, of the review's completion), and have the policy and procedures approved by WorkCover.

Employers must assist in, or provide injured workers with, rehabilitation, which must be of a "suitable standard". A rehabilitation plan must be developed for each worker who is incapacitated from work. Employers with 30 or more workers must also appoint a fully trained rehabilitation coordinator, who must be accredited by WorkCover. Employers must provide injured workers with suitable duties, and workers cannot be dismissed for three months after the injury for reasons relating to the injury.

South Australia

The WorkCover Corporation is responsible for the rehabilitation of injured employees. The corporation is required to establish or provide rehabilitation programs, and can appoint rehabilitation advisers.

Employers must provide suitable employment for their injured employees, unless the employer can show that this is not reasonably practical. Employers with 10 or more employees must keep an injured employee's position open indefinitely, and those with fewer than 10 employees must keep the position open for 12 months.

Western Australia

The relevant legislation in Western Australia "encourages" employer-based rehabilitation programs. The Workers Compensation and Rehabilitation Commission has the authority to require an employer to take reasonable steps to facilitate the rehabilitation of a worker.

Where feasible, employers must provide rehabilitated workers with their original job prior to injury. If that position is no longer available, or if the worker can no longer perform the job, the employer must offer a similar position that the worker is qualified for, capable of doing, and that commands comparable status and pay.

Employers must keep an injured employee's job open for 12 months and must take reasonable steps to rehabilitate the employee.

Tasmania

A rehabilitation policy must be prepared by employers of more than 20 workers and it must be displayed at the workplace.

If a worker is totally or partially incapacitated for more than 14 days, the employer must prepare an RTW plan in consultation with the worker. This must be done within five days of the end of the 14-day period of incapacitation. Where feasible, employers must offer suitable alternative duties that the worker could reasonably be expected to perform. In addition, employers must keep an injured worker's job open for 12 months unless the reason for the original employment no longer exists, or it is not feasible to make that employment available to the worker.

Australian Capital Territory

The employer must provide the worker with occupational rehabilitation at the employer's expense. Employers must develop an occupational rehabilitation policy and review it from time to time. A rehabilitation coordinator must be appointed by the employer. WorkCover ACT administers the legislation.

Northern Territory

Northern Territory employers must bear the reasonable costs incurred for a long-term incapacitated employee's rehabilitation training. This includes a workplace-based RTW program and workplace modification. As far as possible, employers must provide injured workers with, or assist them in finding, suitable employment, and participate in efforts to retrain workers.

Chapter 9

MANAGING RISKS OF ACCIDENT OR INJURY

ALCOHOL AND OTHER DRUGS

FITNESS FOR WORK

DRIVER AND FLEET SAFETY

MANAGING PARTICULAR RISKS

¶9-000 Overview

Chapters 9 to 11 take the form of an introduction to managing some of the more common health and safety problems that exist in Australian workplaces.

Occupational health and safety research into and discussion of each topic is a dynamic area. Since a great deal of specialist research exists on each topic, readers seeking detailed information are advised to consult specific references (limited references are provided at the end of each topic).

One useful source of information that highlights current issues is *The Journal of Occupational Health and Safety — Australia and New Zealand* (CCH Australia Limited). The Journal issues six times each year and publishes writings on subjects of ongoing interest. It draws together current research as well as encouraging wider discussion of the topics featured. Where appropriate, references to issues or articles published in the Journal are given at the end of the sections.

While not all of the problems mentioned in these chapters apply to every workplace, potential risks of these types need to be anticipated and either removed or minimised. Note also that specific types of workplaces may contain other problem areas not discussed in these chapters. Again, consultation of specialist references may be necessary.

The overall aims of these chapters are to acquaint readers with the main issues of each problem, and to outline the types of corrective measures that are possible. The general guidelines provided here are not a substitute for specialist assistance.

The problems described in these chapters include issues that arise from the risk of accidents, risks to health and wellbeing, and social effects and behaviour. Workers compensation systems, however, distinguish between just two categories — occupational diseases and safety problems. They do this on the basis that in diseases the harm is caused gradually (eg hearing loss, back pain or other problems discussed in Chapter 10), while safety problems tend to have immediate and visible consequences (eg hand injuries in machinery, slips and falls or other issues outlined in this chapter). In practice, many problems can have adverse effects on both health and safety, eg shiftwork, or drug or alcohol misuse. Chapter 11 focuses on issues with more prominent social or behavioural elements.

During the risk management process, all risks associated with particular jobs/tasks should be considered in an integrated manner, eg risks associated with the use of a lathe in a joinery should be considered in terms of safety hazards (eg risk of hand injuries or clothing entanglement) as well as health hazards (eg noise and hazardous substances such as wood dust).

Public health issues (eg Legionnaires' disease, or the impact of a lead smelter on the neighbourhood around the smelter) also need to be managed using risk management processes.

WORKING ENVIRONMENT

¶9-100 Introduction

The physical working environment is a fundamental OHS concern. This term covers aspects such as workplace layout including space allocation, ergonomics, lighting, temperature, ventilation and space.

These matters tend to be of a highly technical nature. State legislation sets minimum standards for most of them. Codes of practice and Australian standards may also apply. There may also be extra requirements for certain work environments such as bakeries, foundries, electroplating, abrasive blasting, spray painting and welding. In general, outside professional advice on these topics will be required. The purpose of the following paragraphs is to provide some indication as to whether working conditions are adequate.

Work environments should be carefully considered when new office equipment is being installed and should be considered in consultation with a workplace health and safety committee and/or representative. A common example has been the introduction of computers. These involve a great deal of close eye work and constrained body posture, and so require favourable lighting and ergonomic layout in particular.

Finally, although technical advice may be needed, do not hesitate to seek consultation with and suggestions from health and safety representatives or employees where a committee, or equivalent, is not established. The workers are the ones who have to cope with and adapt to the conditions provided, so it is appropriate that they have some say in the matter; in any case, a consultative approach is fundamental to managing health and safety. Employees may also be able to alert management to potential problems and use their on-the-job knowledge to come up with valuable suggestions.

¶9-120 Workplace layout

It is not possible to set out guidelines that apply to every workplace, because of the tremendous differences between work environments, dictated by type of job, equipment, and so on. However, the following summary provides some idea of which aspects to consider.

Summary — workplace layout

1. Flexibility is preferable — so that the layout can be easily changed as needs change. For example, avoid using permanent room/desk dividers when movable ones will be equally effective.
2. Layout should facilitate communication between employees, so that work flow is assisted. People whose jobs are interrelated should, where possible, be positioned near each other.
3. A certain amount of privacy will also be required. For example, a closed room for interviews in the human resources department may be needed. Also, employees require some protection from possible sources of distraction. Work requiring concentration and close attention can easily be compromised in overcrowded, noisy work environments.
4. Employees should have adequate working space, as well as furniture, storage space, etc. Space allocations per worker are not specified in OHS legislation, but are mentioned in some other guidance material. As a rule of thumb, it is generally recommended that 6.25 square metres should be allocated for each individual workstation, that is, including furniture and fittings, but not including passageways and amenities.
5. Workstations should be designed so that employees are not repeatedly required to bend, stretch, twist their torsos or raise their arms above their shoulders. Commonly used controls, switches and tools should be placed within comfortable reach of the employee.
6. Some effort should be made to provide a pleasant office or factory landscape (for example, through use of colours, position of windows, posters, etc).
7. Some recreation space and adequate passageway space should be provided.
8. Facilities such as toilets, locker rooms, first aid room, tearoom, etc, should be adequately equipped and readily accessible.
9. Direction signs (for example, location of departments, names on doors) should be clearly and prominently installed.
10. Equipment such as air-conditioning, heating, switch room, etc, should be positioned so that its operation (noise level, etc) will not cause discomfort to any employees.
11. Safe means of entry and exit from premises is required, as well as rapid exit means in case of emergency.

¶9-130 Ventilation

State regulations (both safety regulations and building regulations) cover aspects of ventilation and extraction. Fresh air, supplied by either natural or mechanical means, is needed to keep down the level of carbon dioxide and to dissipate odours and contaminants.

The three main types of ventilation are natural, dilution and local exhaust ventilation.

- *Natural ventilation* is provided by outdoor air movement. Size and positioning of doors and windows is important. Natural ventilation cannot be relied upon to remove contaminants from the atmosphere.
- *Dilution ventilation* refers to the removal of old air and the replacement of it with clean or "make-up" air. It generally relies on exhaust fans

removing old air from the building. Dilution ventilation is suitable where there is a low level of contamination, especially if people do not work near the source.

- *Local exhaust ventilation* works by capturing the contaminant near the point of release and removing it from the operator's breathing zone. Local exhaust systems are necessary in the case of toxic gases or fumes, or where significant quantities of dust are generated. Common types include extraction fans and exhaust hoods.

Exposure limits, which set out safe levels of contaminants in the atmosphere, are discussed at ¶10-230.

Ventilation for climate control

Air-conditioning is a common feature of indoor workplaces. State regulations generally cover the construction and maintenance of air-conditioning systems. Systems should be positioned to minimise noise level problems. Maintenance must be adequate to check for problems such as Legionnaires' disease, which is caused by a bacteria that can grow in the water of air-conditioning systems.

Further references

Industrial Ventilation: a Manual of Recommended Practice, American Conference of Governmental Industrial Hygienists, 23rd edn, Cincinnati, Ohio, 1998.

The Hands On Guide: OHS Legal Guide, CCH Australia Limited, 1998.

The Hands On Guide: OHS Manager, CCH Australia Limited, 1998.

Toxic Gas Generated by Micro-organisms in Confined Spaces, WorkCover New South Wales, 2003.

Safetyline Essentials: Local Exhaust Ventilation, WorkSafe Western Australia, 1999.

Managing Indoor Air Quality, Building Owners and Managers Association of Australia Limited, Sydney, 1994.

Air-conditioning and Thermal Comfort in Australian Public Service Offices, Comcare Australia, 1994.

Code of Practice: Prevention and Control of Legionnaires' Disease, WorkSafe Western Australia Commission.

Officewise: a Guide to Health and Safety in the Office, Victorian WorkCover Authority, 1997.

Indoor Air Quality, Workplace Health and Safety, Queensland, 2002.

Work Environment — Indoor Air Quality in Air-conditioned Offices, NT WorkSafe, Northern Territory, 2001.

AS 1668.2-2002: *The use of ventilation and air-conditioning in buildings — Ventilation design for indoor air containment control*, Standards Australia.

¶9-140 Temperature

Discomfort at work can be caused by temperatures that are in the extreme ranges of heat or cold. Each work activity has a temperature range in which it is usually comfortable to perform the task. Temperature measurements are ideally taken at individual workstations, 1200 mm above floor level.

State regulations may set minimum temperature limits below which radiant heating may be required to maintain temperature levels. Some occupations, however, can only rely upon warm clothing to ensure comfortable body heat (for example, cool store or refrigeration workers and outdoor workers in cold climates).

In buildings where the work carried out generates large quantities of water vapour (steam or spray), such as laundry pressing rooms and kitchens, the water vapour should be collected at the source and removed by the use of suitable enclosures, hoods and extraction systems.

In cases of high temperature, the following actions can be taken:

- introduce additional mechanical ventilation with suitable temperature control devices
- increase air movement within the workplace
- install evaporative coolers or air-conditioning
- reduce entry of solar heat to building by, for example, placing reflective material on windows, awnings, shades and blinds
- insulate heat-generating locations with barriers and screens, and use extraction systems to remove hot air and vapours
- reduce the periods of time for which workers are exposed to extreme conditions
- in extreme conditions, provide special protective clothing with in-built cooling systems, and
- provide cool water dispensers.

A simple increase in the air flow is effective in cooling the workplace where the air temperature is below 35°C. In conditions that are not too humid, evaporative coolers can be effective. In higher temperatures with humid conditions, only air-conditioning will be effective in reducing the temperature, although other measures (such as increased air movement, awnings) can have some effect.

Further references

ACTU Guidelines for Working in Seasonal Heat, Australian Council of Trade Unions OHS Unit, Victoria, 2002.

Working in Heat, Victorian WorkCover Authority, Melbourne, 1998. (See WorkCover's website at www.workcover.vic.gov.au)

Air-conditioning and Thermal Comfort in Australian Public Service Offices, Comcare Australia, 1994.

Heat-related Illness: Signs, Treatment and Prevention, Workplace Health and Safety, Queensland (undated).

¶9-150 Lighting

Lighting is another highly technical subject, also covered by state regulations, codes of practice and Australian standards. The following guidelines will help provide employers with a useful introduction to the subject.

GUIDELINES — LIGHTING

1. Work area illumination should be suitable for all intended purposes, and reasonably uniform throughout. If a higher level is needed for particular tasks, the overall level should be increased to reduce contrast. Brightness distribution should be controlled.
2. Lighting should be installed so as not to cause glare within employees' fields of vision.
3. Colour. A white or near-white non-glossy finish on the ceiling or roof underside is preferred. Reasonably light colours are preferable for other surfaces. Remember that colours will greatly influence the lighting effect and employees will be subjected to these conditions during their full period at work.
4. Some access to natural daylight (windows, etc) is both necessary and desirable. It should be provided in conjunction with artificial lighting.
5. Alternative emergency lighting facilities should be provided in case of failure of the normal system.
6. Areas such as stairwells and escape exits should be illuminated and remain lit whenever the building is occupied.
7. Individual tasks may require special study, such as bad reflections, close visual concentration, positioning of employee, details reflected on shiny surfaces, etc. A common example is reflections on the screen of computer monitors or other visual displays. Careful planning of office layout will be needed to overcome these difficulties.
8. Faults, such as flickering lights, blown globes, etc, should be attended to as soon as possible.
9. Sometimes a stroboscopic effect may occur when revolving machinery parts are located in proximity to light fittings. Bays of lights may remove this effect.

Further references

Health and Safety in the Office, WorkCover NSW, 1993.

Officewise: a Guide to Health and Safety in the Office, Victorian WorkCover Authority, 1997.

Artificial Light at Work (1984), *Colour at Work* (1980) and *Daylight at Work* (1983), Department of Employment and Industrial Relations, now published by the National Occupational Health and Safety Commission.

AS 1680: Interior Lighting series, Standards Australia.

See also the websites of government OHS authorities and other OHS information providers such as those listed in the Appendices.

INJURIES FROM PLANT

¶9-220 Scope of plant/equipment injuries

The risk of injuries from machinery and equipment has long been recognised as a key focus of employers' duty of care. "Plant" is a term that includes "any machinery, equipment (including scaffolding), appliance, implement or tool and any component or fitting thereof or accessory thereto". Plant injuries include injuries involving means of transport (cars, trucks, trains), traditional industrial machinery (lathes, manufacturing equipment, printing presses), lifting equipment (cranes, fork-lifts) and hand tools (power saws, staple guns). Injuries from these sources account for over 40% of all workplace injuries and fatalities.

¶9-230 Purchasing

Good engineering can provide permanent solutions to problems with plant safety. Purchasing specifications must incorporate all required safety features (such as adequate guarding of dangerous moving parts). Where sufficient expertise is not available within the organisation, it may be necessary to engage engineering and/or ergonomic consultants.

Advice from operators and maintenance staff should be sought prior to purchasing equipment. These are the people most likely to know the safety features which should be incorporated into any new machinery.

Part of any acceptance check of new equipment should be to determine that the equipment meets the safety specifications of the organisation, in addition to legislative requirements concerning plant safety or machine guarding.

Legislative requirements affecting manufacturers and suppliers

Health and safety legislation in each state and territory places a duty on designers, manufacturers, suppliers, erectors and installers of plant to ensure the safety of plant that they provide. This includes the adequate guarding of any dangerous parts. It may also include a requirement to carry out safety testing on the plant, and to supply information relating to the plant.

Suppliers of defective plant that leads to an injury at the workplace could find themselves liable to a fine under health and safety legislation, as well as a common law action based on negligence.

¶9-240 Machine guarding

Where a hazard is associated with plant, it should be eliminated or isolated. When this is not possible, however, guarding may be the next preferred option. Guards need to be provided whenever there is the potential for people to come into contact with hazardous situations. So, for example, guards are required to cover very hot pipelines or to protect people's eyes during grinding processes. Guards must be provided whenever there is the possibility of a person's hands, hair, clothing, etc, getting caught in or coming into contact with moving machinery. Guards should be provided whenever there is the possibility of objects flying off moving machinery.

Some common machinery parts that require guarding include:

- flywheels connected to a moving engine
- gearing equipment
- belt and pulley drives
- chain and sprocket drives
- electric generators
- motors
- feed-in rollers
- exposed electrical conductors, and
- rotating or reciprocating machine elements.

Design of guards

Where possible fixed guards should be used. These are guards that do not need to be removed for inspections and routine servicing. Where it is necessary to remove a guard for servicing, it should be secured so that it cannot be removed without the use of tools.

Guards should be sufficiently strong to support the weight of a person standing or leaning on them. They should have no sharp edges or protruding corners.

Movable guards must be interlocked with the machine, either mechanically or electrically. This means that the machine cannot be started, or cannot continue in operation, without the guard being securely in place. The guard should not be able to be opened while the machine is in operation. Where electrical interlocking is used, it should be "fail safe" (so that the machine will shut down if something goes wrong).

Another alternative where access to the dangerous parts is required is to install a trip device (such as photoelectric beam or pressure sensitive mat). The machinery will shut down when the trip device is activated.

Legislation covering machine guarding

In each state and territory the employer is under a general duty to provide a safe workplace for employees, including safe plant (see ¶2-130). Historically, there has been a legislative focus on guarding all dangerous machinery and/or setting out guarding requirements for specific machinery. The machinery that is likely to be subject to specific requirements includes:

- chaff-cutting machines
- circular saws and other wood-cutting machinery
- explosive-powered tools
- foundry equipment
- power presses

- steam pipes, and
- welding equipment.

There are also Australian standards dealing with the guarding of specific machinery (see "Further references" at the end of ¶9-260).

The legislation relating to guarding is tending to be replaced with new legislation based on the risk management approach (see Chapter 3) and consistent with the national standards described below.

National standards for plant and operator certification

Two national standards target health and safety related to plant and equipment use. These are the *National Standard for Plant* and the *National Occupational Health and Safety Certification for Users and Operators of Industrial Equipment*.

The *National Standard for Plant* sets out hazard identification, risk assessment, and risk control procedures for all types of plant. The certification standard specifies core competencies which are held to be the basic requirements for safe equipment use and operation. Now that it has been adopted in all states, the national certification system, with certificates issued in one jurisdiction recognised by all others, is in place.

The two standards are complementary, so that, for example, anyone undertaking work in conformance with the plant standard, such as the erection and installation of particular plant, is required to hold certification in accordance with the certification standard.

Legislation covering plant and certification of operators

Specific regulations covering plant and certification of operators apply in most jurisdictions. Some of the regulations adopt the national plant and operator certification standards declared by the National Occupational Health and Safety Commission.

The common essential requirements of the national standard have been adopted by all states and territories. Most jurisdictions have incorporated the requirements as legislation, but Queensland has developed the *Plant Advisory Standard*, and the Australian Capital Territory has adopted the *National Standard for Plant* as a code of practice.

¶9-250 Maintenance

Adequate maintenance keeps plant running productively, as well as safely.

Preventive maintenance versus breakdown maintenance

There are two possible strategies regarding the maintenance of machinery:

1. Run the machinery until it breaks down, then either repair or replace it. This is known as breakdown maintenance (also on-call, unplanned or band-aid maintenance).

2. Develop a planned series of tasks, based on the likely service life of the machinery, that will keep the machine in working order. This is known as preventive maintenance.

Choosing between the two systems of maintenance is a management decision. However, this decision should not be guided purely by economic factors. The time loss associated with a machinery breakdown, as well as the possibility of injury to employees and damage to other equipment, should be considered.

Preventive maintenance

Preventive maintenance, by its nature, requires a regular pattern of tasks to be performed. It lends itself to the preparation of written procedures and/or checklists. A maintenance procedures manual should accompany every major machinery purchase.

In some cases, preventive maintenance will be required by legislation. Examples include boilers and pressure vessels, lifts and hoists, and air-conditioning water coolers.

Increased risks to maintenance workers

Maintenance workers will often be exposed to a higher safety risk than other workers. They will be working on or within the containment systems designed to protect other workers.

For this reason it is important to have rigidly adhered to safety procedures in place for maintenance workers. These may include a system of isolation or lockout (see ¶9-260). Where maintenance workers are working in areas containing dangerous chemicals, or in hot or confined areas, it may be necessary to monitor their wellbeing closely.

Where contract maintenance is used, it will be important that the maintenance workers know the organisation's relevant safety procedures. These could include the location of first aid facilities, fire and emergency procedures and local lockout systems.

¶9-260 Danger tags and lockout systems

During the installation of new plant or maintenance of existing equipment, a lockout safety device system can be designed and incorporated into the main control mechanisms.

The lockout safety device system requires each person to be issued with their own lock and key. The keys are not interchangeable. The employee places the lock on the lockout device that is fitted to the main control mechanism and it remains there until the work is completed. The employee alone must remove the lock. Strict observance of this rule prevents accidents to those working in out-of-sight locations.

On small plant and equipment it is possible to install a lockout device or lockout main switch in the original installation, so it can be locked in the "off" position. This method is more positive than just the removal of fuses or an iron bar through the hand wheel of a valve.

A danger tag system does not offer as positive a means as the previously mentioned isolating devices, but it is designed to offer a measure of personal protection on existing plant and equipment without major capital outlay for modifications. Also, the danger tag system can be effectively introduced to promote greater safety consciousness amongst the workforce and does offer a start in providing a definite hazard spotting system within the working procedures.

An example of tags is shown below.

The danger tag system should embrace all situations in which danger to persons could arise from the operation of machinery, plant or equipment, by the flow of steam, electricity, gases or liquids, or the use of faulty or unsafe plant and equipment. As a bonus, the system can minimise further damage to plant and equipment caused by inadvertent starting.

Danger tags have a wide range of applications, including large machinery and equipment in factories, "field" sites, and small items of plant and equipment, such as tools, ladders or defective vehicles (in the latter case, the tag could be attached to the steering wheel).

The system could embrace a control return point where defective portable items can be returned for repair to a workshop, thus speeding up the procedural flow and minimising the time that the unit is out of service. This allows plant and equipment to be used to its fullest degree, provides for maximum efficiency and minimum duplication of units.

Further references

Plant Safety: Managing Plant Hazards in the Workplace, CCH Australia Limited, 1997.

The Hands On Guide: OHS Manager, CCH Australia Limited, 1998.

Forklift Safety — Reducing the Risk — Forklift Instability, WorkSafe Victoria, June 2005.

Plant in the Workplace: Making it Safe, National Occupational Health and Safety Commission, 1995.

Plant Design: Making it Safe, National Occupational Health and Safety Commission, 1995.

National Occupational Health and Safety Certification Standard for Users and Operators of Industrial Equipment, National Occupational Health and Safety Commission, 3rd edn, 2001.

National Standard for Plant, National Occupational Health and Safety Commission, 1995.

Controlling Risks from Plant (Machinery and Equipment), ACTU, 1996.

Standards Australia publishes a range of standards on the guarding of specific machinery, including: AS 1893 (guillotines), AS 1473 (woodworking machinery), AS 1788 (abrasive wheels), AS 2153 (agricultural tractors and machinery), AS 1873 (explosive-powered fastening tools), AS 1219 (metal power presses), and AS 4024 (safeguarding of machinery).

See also the websites of government OHS authorities and other OHS information providers such as those listed in the Appendices.

SLIPS, TRIPS AND FALLS

¶9-320 Scope of the problem

In the workplace context, "slips, trips and falls" is a general term referring to accidents involving:

- falls on the same level, and
- stepping on an object.

It does not include falls from a height, which are covered separately at the end of this section (see ¶9-350).

The falls resulting from slips and trips result in a wide range of injuries including back strains as well as injuries to other parts of the body. Incidents of this type in Australian workplaces are major causes of permanent disability. These injuries account for around 19% of the total number of serious injuries to working people. Falls on the same level are responsible for nearly 11% of back injuries, and those injuries are likely to be more severe than those resulting from over-exertion.

The average cost of compensating a slip, trip and fall accident is around $10,000, and businesses can incur additional costs in recruiting and training a new worker while the injured worker recovers, and in lowered productivity.

If the reported instances of non-employees injured in workplaces were to be added to the employee statistics (one large retailer, for example, reported that about 80% of its customer injury claims are associated with slip and fall accidents), the losses resulting from slip and trip injuries would be even larger.

¶9-330 Causes of slips, trips and falls

Slips occur when a person loses their footing or balance when their feet unintentionally slide over the surface they are in contact with. Slips may also include instances where other bodily parts unintentionally slide on some surface that is bearing body weight (such as the fingers losing their grip on the rung of a ladder).

Trips occur when a person loses their footing or balance when they unintentionally stumble (often after catching their foot on some object that momentarily prevents their foot from moving in the direction in which the body weight is being transferred).

Slips and trips can occur in many different ways. Slips in the workplace result from circumstances such as:

- surfaces having poor friction characteristics (particularly when they are wet)
- surfaces being dirty, or having some slippery substances or material on them, such as lubricating oil, foodstuffs and beverages, product, fine dusts, water and rollable solids (such as beads, pellets, spherical ball bearings, pipes and regular-shaped objects)
- people wearing unsuitable footwear such as shoes with soles that have no patterned grip, are worn smooth, or have poor frictional characteristics
- surfaces being inclined or irregular
- people walking on surfaces where there are no handrails or balustrades, and
- people running on polished or smooth (low friction) surfaces.

Trips in the workplace result from circumstances such as:

- floor surfaces being rough or uneven
- floor coverings, particularly carpet or matting, being torn or uneven
- hazards in which people may catch the toe or heel of their footwear
- some obstruction (such as trailing electrical cables or objects left on the floor)
- stairs having uneven "riser" or "going" dimensions, or stairs that fall outside the parameters of safe design (as established in AS 1657-1992)
- variations in floor levels that are difficult for people to appreciate or perceive (often because of poor lighting), and
- people wearing inappropriate footwear (such as high heels on grid-mesh flooring).

¶9-340 Managing slip and trip hazards

Most Australian OHS legislation has requirements for the design, installation and maintenance of floors, passageways and stairs that existed prior to risk management-based provisions. These older provisions require the owner or occupier to meet prescribed standards. Several OHS statutes have now called up standards such as AS 1657-1992: *Fixed platforms, walkways, stairways and ladders — Design, construction and installation*. Employers and occupiers bear the main obligations, although the designer also has certain responsibilities.

Aside from any specific or prescriptive requirements, all employers are under a general duty to provide a safe place of work, and this broad obligation may be taken to include the duty to prevent injury resulting from slip or trip hazards.

The primary means of achieving this is to undertake risk management in relation to all slip and trip hazards in the workplace — that is, to:

- identify all slip and trip hazards
- assess the associated risks, and
- control those risks.

There is also a need to ensure that sufficient monitoring and review is undertaken of risk management systems to ensure they remain effective, and to keep records of the risk management process (see Chapter 3). Further details on managing slip and trip hazards are provided below.

Principles for controlling risks from slips and trips

There are many ways in which the risks associated with slip and trip hazards can be controlled. These include:

- the use of high grip floor coverings
- provision of collection points for liquid spills, and
- the enclosure of machine parts that generate dust, oil or other waste substances, so that any spillage or leakage is retained.

The elimination of slip and trip hazards must be a primary aim of workplace and plant designers and those in control of workplaces. The risk of injury associated with slip and trip hazards can be effectively minimised by:

- selecting/specifying floor surface and paving materials that give good frictional grip
- ensuring that any wet areas where liquid may accumulate will drain to a collection point. A commonly encountered problem with wet floor areas is that they are laid flat, leading to liquid "ponding" on the floor rather than flowing away from trafficable areas
- enclosing equipment that generates dust, oil or any other substances that may foul trafficable surfaces
- locating sources of dust, oil and other substances that may present slip or trip hazards in a bounded area
- designing all stairs and ladders so that they fully conform with the requirements of AS 1657-1992
- avoiding variation in floor levels
- providing clear accessways that are free of obstructions and objects protruding into walking zones
- clearly delineating the designed accessways
- providing adequate lighting (including emergency lighting) over areas where people are required to walk

- providing suitable handrails and balustrades, and
- using techniques such as "tiger taping" to make any unavoidable differences in floor level as visible and obvious as possible.

Manufacturers' duties

The manufacturer of plant items can play a role in ensuring that slip and trip hazards are not built in as part of the manufacturing or assembly process. Flooring, in particular, should be fitted and finished well to avoid any inconsistency or unevenness. Cabling and piping should be secured so that it cannot accidentally snag feet in tight spaces (such as in the operator's cab of mobile plant).

Installers' and erectors' duties

The installer/erector of plant must ensure that during the construction, erection, installation and commissioning stages, slip and trip hazards are identified and either eliminated or, at least temporarily, isolated (for example, by the erection of barricades or fencing). Where no specification for floor finishes in pedestrian areas is given, the opportunity may exist to introduce additives to floor finishes (such as carborundum) to increase the frictional grip of the flooring.

Owners' and occupiers' duties

The owner or occupier of premises where plant is installed must identify slip and trip hazards as part of the risk management process for the plant. The ongoing control measures adopted should include regular inspections of the workplace by local management to identify any slip or trip hazard that may have been overlooked in the initial risk management process, or that has been introduced since the last inspection.

¶9-350 Preventing falls from heights

Working at heights involves those situations where people work in a position which is elevated above a local floor or ground level, or which is close to the edge of a drop. In addition to those undertaking tasks such as roofing, electrical lines work, building construction, scaffolding and rigging, people using hoists, cherry pickers and ladders are potentially at risk of falling.

As well as being covered by the general requirement to have a safe system of work, such work situations may also be subject to particular legislative provisions requiring precautions to be taken to avoid the risk of people falling some nominated distance (for example, 1.8 metres). Some work at heights, such as roofing, may also be covered by codes of practice. In any case, the risks inherent in the work need to be identified, assessed and controlled.

Safety precautions for working at heights

Accidents and injuries when working at heights may be avoided by adopting the following precautions:

- ensuring that properly erected and maintained access and egress arrangements to and from raised work areas are provided
- operating plant and equipment used for working at heights within the safe envelope specified by the manufacturer, designer or supplier
- maintaining plant and equipment (and especially load shifting equipment) used for working at heights within the safe envelope specified by the manufacturer, designer or supplier
- providing adequate working platforms, with secured access ladders, guard rails and toe-boards
- ensuring that crawling ladders or boards are secured in position and capable of bearing the required load
- ensuring that access on or over fragile materials is guarded or that any fragile material is suitably covered
- displaying prominent warning notices at all access points to fragile materials
- ensuring that safety nets are properly installed, maintained, inspected and tested, and that they are as close as possible below the work
- ensuring that safety harnesses, where worn, are securely anchored and that they are properly maintained and inspected
- ensuring that risk management processes are carried out in consultation with people who are required to undertake work at heights (and their representatives), and that the results of these processes are reviewed regularly
- informing workers of the results of risk management processes and the measures which will be taken to reduce risks related to working at heights
- ensuring that work at heights is carried out under competent supervision
- providing appropriate training to all personnel involved in working at heights.

Further information

Basic Steps to Preventing Falls from Heights, WorkSafe Victoria, June 2005.

Guide to Preventing Slips, Trips and Falls, Comcare Australia, 2002.

Significant Incident Summary: Slips, Trips and Falls: Hospitality Industry, WorkSafe Western Australia. (Website at www.safetyline.wa.gov.au/PageBin/injrsign0195.htm)

Slips, Trips and Falls, WorkCover Corporation of South Australia. (Website at www.workcover.com)

Falls in the Workplace, Safetycare Australia, video, 16 minutes, 2002.

Preventing Slips, Trips and Falls — guidance note, WorkCover NSW, November 1998.

Avoiding Slips, Trips and Falls — information for workers, WorkCover NSW, November 1998.

See also the websites of government OHS authorities and other OHS information providers such as those listed in the Appendices.

EMERGENCY PREPAREDNESS

¶9-400 Need for emergency planning

There is always a possibility that some chain of events at a workplace or involving work activities, or some event arising from outside the workplace, could lead to an emergency situation. An emergency is a situation that is dangerous, or potentially dangerous, to life, property or the environment.

Emergencies are generally considered to involve an abnormal and sudden occurrence. They may result from an event such as:

- a leak or spill
- a fire or explosion involving hazardous materials
- a transportation accident, or
- an uncontrolled release (or potential release) of damaging energy resulting from events such as
  - the mechanical failure of a machine element
  - the overloading or failure of an electrical conductor, and
  - the emission of harmful radiation.

In these days of heightened awareness of the possibility of terrorism, it may be considered appropriate to include a bomb threat or terrorist attack in the list of contingencies for which the organisation can make suitable preparations.

Emergency response procedures outline the required actions and arrangements that have been devised and maintained to deal with an emergency situation.

¶9-420 Emergency response plans

Employers should have contingency procedures in place to deal with emergencies. These are designed to prevent, contain or minimise any potential injury or damage. This contingency plan is usually referred to as the "emergency response plan", and is a formal document outlining the:

- types of emergency scenarios which can be reasonably expected
- chain of command that will be adopted during the emergency (including the duties of key personnel)
- systems and procedures to be activated in order to cope with an emergency situation
- administrative arrangements to support the emergency preparedness of the site (including training, debriefing, auditing, and practice schedules), and

- business continuity plans for activation in the event that operations at the site are severely disrupted.

The development of emergency response plans, and the process of reducing the potential risk of plant failure or other emergency conditions, should be systematic and well structured.

What is an emergency response plan?

In order to ensure that people are protected from risks associated with an emergency situation, it is necessary to have a site evacuation procedure. In developing this procedure, the following elements must be considered and appropriate schemes devised to deal with:

- the arrangements for raising an alarm
- accounting for people
- escape routes and muster points, and
- the declaration of the end of the emergency.

It is an unfortunate misconception, but many organisations regard their workplace evacuation plan as being an adequate emergency response plan. While the existence of an evacuation plan is an essential component of an emergency response plan, there are other issues that should be included in any comprehensive plan.

Elements of the emergency response plan

The key elements involved in devising an appropriate emergency response plan are as follows:

- the identification of particular hazardous inventories or tasks, and the description of any safety-critical control measures designed to avoid or minimise escalation of an emergency, or to contain damage
- responses other than evacuation (for example, spill containment and clean-up, damage control, firefighting and rescue arrangements)
- relevant training needs, eg the need to train specific personnel in how to respond to a bomb threat
- arrangements for the reporting of incidents to statutory authorities
- arrangements for media liaison
- procedures for restoration and rehabilitation
- arrangements for contacting key personnel (both during normal working hours and after hours) and support organisations (such as suppliers of goods and services likely to be required, counsellors, and organisations which can assist with off-site matters)
- emergency equipment schedules, and operating and maintenance instructions for emergency equipment, and
- a business continuity plan.

External emergency services

The resources needed to combat an emergency often exceed the resources immediately available at the site. This makes most organisations very reliant on emergency services (such as the police, fire brigade, bush fire brigade, ambulance and port authorities) for assistance in coping with or combating an emergency. The emergency response plan needs to describe the procedures for summoning outside assistance and the arrangements for liaising with the external emergency services combat team.

Emergency services personnel require as much information as possible about the nature of the hazard they face, and particularly any information which will help them to assess what is likely to happen if the emergency cannot be quickly controlled. The organisation needs to have arrangements in place to provide data such as the emergency services manifest (an inventory of all dangerous goods on the premises, or in the vehicle, itemised by class and location).

¶9-430 Key requirements for emergency response contingency planning

Emergency response planning is an integral part of occupational health and safety management (falling within the general duty of care requirements to provide safe systems of work and to make arrangements such that nobody's health or safety is at risk due to workplace activities).

The principal occupational health and safety statutes make no distinction or allowance, when outlining accountabilities, between normal and abnormal operating modes. All of the following have some accountability for ensuring the adequacy of the emergency response arrangements that are in place and the efficient implementation or operation of those procedures:

- designers and manufacturers
- importers and/or suppliers
- installers or erectors
- owners or operators
- employees, and
- others.

Assignment of duties

The basic requirement for being able to competently deal with any emergency situation is to be adequately prepared and, so far as is possible, efficiently organised.

The task of devising and/or maintaining an emergency response plan is often undertaken by either a functional manager (such as the safety manager or the personnel manager) or may be outsourced to a specialist consultant who is competent in emergency response matters.

Considerable local input in the workplace is necessary to identify hazards specific to particular areas and to select appropriate emergency organisation structures, hardware and procedures which will function adequately for the organisation. When developing or reviewing an emergency plan, senior line managers, people assigned specific roles under the plan (such as fire wardens and first aiders) and members of the organisation's occupational health and safety committee should all be involved.

Key in-house emergency personnel

The emergency response plan should identify the key roles that must be undertaken, clearly stating the principal and stand-by personnel in those roles. It is usual to set up positions such as:

- the emergency controller
- evacuation wardens
- duty first aiders
- the services and utilities coordinator (to isolate/operate key emergency services and site utilities such as fire, water, electrical power and pollution prevention devices)
- personnel responsible for checking emergency equipment for serviceability, accessibility, cleanliness and correct location
- the media relations and reporting coordinator, and
- first aid, fire and rescue teams.

In addition, staff likely to receive phone calls conveying bomb threats or other such communications should be trained in how to respond.

Finally, the plan should identify who is responsible for:

- auditing the emergency response plan (by routine assessment and through incident debriefing)
- maintaining the plan, and
- if necessary, reviewing and upgrading the plan.

¶9-440 Principles for devising and maintaining an emergency response plan

The first step in developing an emergency plan is to identify foreseeable emergency situations, which may include:

- fire and explosion
- toxic release
- bomb or extortion threats
- medical emergencies
- environmental damage threats (such as a spill or release), and
- external events threatening the site (including natural disasters).

The likelihood of a terrorist attack is difficult for most Australians to contemplate under the heading "foreseeable", nevertheless it may be appropriate to consider the possibilities. It should be remembered that targets favoured by terrorists include places where large numbers of people congregate, such as theatres, schools or clubs, and places which may be vulnerable for political reasons, such as embassies. Even when the organisation does not come into any of these categories, its physical location may increase the risk, due to the proximity of other premises, which could conceivably be targets. The presence of dangerous goods, for example materials that could be used to make bombs, may also raise the chances of some type of attack or threat.

Developing suitable response systems

Having identified what can go wrong, the next step is to develop suitable responses which are necessary to manage the consequences of those events (or combinations of events) which are considered to be at risk of occurring. This process should take account of relevant standards, current industry practice and specialist advice. It is necessary to consider:

- the internal organisation arrangements for the emergency response team (including positions, people and duties)
- the procedures for responding to the various types of scenarios identified (procedures should be documented and displayed)
- how people may be provided with a temporary safe refuge during an emergency event, and how to safely evacuate a site in the event that this may be necessary
- how emergency combat personnel will identify and gain access to
  - all service isolation points
  - the installed emergency response equipment (such as fire hydrant points, emergency water supplies, fire booster system connectors)
  - information needed to effectively combat the emergency, and to make sensible decisions on actions to be taken to reduce the threat to people and the environment
- what means of warning/alerting people that an emergency situation has developed should be provided and what (if any) additional signal(s) should be given that an evacuation is to proceed
- what systems for communicating with all personnel in the event of an emergency can be secured and maintained
- what operating procedures should be devised to initiate/deploy specific emergency response equipment (such as an automatic sprinkler system, direct brigade alarm, deluge or drencher system, duress alarm)
- what arrangements should be made for clean-up and restoration works
- what systems are required to assist emergency combat organisations such as the police and fire brigade — for example, the emergency services manifest and a copy of the site emergency plan (both are usually provided at a point of access to the premises)

- what training (and which training providers) might be suitable to prepare employees in terms of what they might expect to experience under various emergency scenarios
- what type of follow-up, debriefing, and assessment of needs such as post-traumatic stress management, might be appropriate in a range of contingencies, and
- what arrangements should be made to ensure business continuity.

In relation to the last point, it has been standard practice in well-prepared organisations to ensure that back-up of critical computer systems and data is maintained off site. As a consequence of the events of September 11, 2001, it has been proposed in the United States that recovery sites should be at least 350 km away from the main site of operations, and critical organisations are being urged to split centralised workplaces into two locations. Instead of having a centralised function with a non-staffed back-up site, it is now being recommended that organisations staff the alternative site as if it were a primary site. The effect would be to have half the staff in one location and half in another. This has long been used as a risk management strategy in other contexts, for example, not transporting the whole football team in a single aircraft.

Duty to undertake risk management

A number of sites are under a specific obligation to devise and maintain emergency plans. These generally include either of the following:

- sites where threshold levels of dangerous goods have been exceeded, and
- sites which meet the classification requirements for a "major hazard facility".

A larger number of sites have devised emergency plans as part of the "due diligence" management process in response to their general duties of care.

There is still no specific comprehensive mandatory requirement in all Australian OHS statutes for conducting the risk management process in relation to the adequacy of emergency arrangements at most existing premises. Some OHS statutes have exercisable powers in relation to dangerous goods and/or planning and development (where hazard analysis and risk assessment may be required). It can, however, be reasonably anticipated that such a requirement will be introduced in all Australian OHS laws over the next several years (at least for sites that are identified as being potentially hazardous).

Irrespective of any legal requirement, having an effective emergency response plan is both an important core element of occupational health and safety management and a basic commercial and business imperative. Failure to have an effective plan, or worse, any plan at all, leaves a very large risk exposure of sustaining catastrophic and irrecoverable damage to the organisation's assets (of people, plant, property and the environment).

Bomb threat

In the case of a bomb threat, immediate evacuation of the building(s) may be necessary, depending on the specific details of the threat. Evacuation procedures should already have been established, and staff trained and drilled in them.

At the same time as protecting staff, contractors, visitors or others on the site, it is also necessary to try to establish the credibility of the threat and, if possible, to collect vital information about the source of the threat.

This means that in the case of a written threat, the note or letter in which the threat was made should be kept, and so should any envelope or container in which it was received. Unnecessary handling should be avoided and every attempt made to retain possible evidence such as fingerprints, handwriting, paper and postmarks.

When the threat is made by telephone, the person receiving the call should not disconnect the caller, and should try to find out relevant information, such as the location of the bomb and expected time of explosion. In addition, they should note details of the caller's voice, speech, background noises and the time of the call.

Details of bomb threats should be reported immediately to security and emergency services, and evacuations initiated as appropriate.

Further references

Barnes, JC, *A Guide to Business Continuity Planning*, John Wiley and Sons, US, 2001.

Emergency Management Practices, Emergency Management Australia. (Website at www.ema.gov.au)

Hiles, A, *Business Continuity: Best Practices*, Rothstein Associates, US, 2000.

Guidance Note for Emergency Services Manifests, National Occupational Health and Safety Commission [NOHSC:3010(1990)], Sydney, 1990.

Hazardous Industry Planning Advisory Papers: Industry Emergency Planning Guidelines, New South Wales Department of Urban Affairs and Planning, Sydney, 1993.

Rothstein, PJ, *Disaster Recovery Testing: Exercising Your Contingency Plan*, Rothstein Associates, US, 1995.

How to Prepare for Workplace Emergencies, US Department of Labor, Occupational Safety and Health Administration, 1995.

Emergency Preparedness for Business, US National Institute for Occupational Safety and Health, Centers for Disease Control and Prevention. (Website at www.cdc.gov/niosh/topics/prepared)

See also the websites of government OHS authorities and other OHS information providers such as those listed in the Appendices.

FIRE PREVENTION AND CONTROL

¶9-520 Aspects of fire prevention and control

The impact of a fire on an organisation can be enormous. Lives can be lost, property and stock damaged or ruined, sales lost and business disrupted while rebuilding takes place. In some cases, the business may even be unable to recommence operations. For these reasons it is in all employees' interests to observe fire prevention and control measures.

Proper provision for fire prevention and control must include each of the following aspects, all of which are outlined later in this section:

- observance of legislation (Acts, regulations and ordinances) relevant to fire prevention and control
- attention to workplace "housekeeping" to reduce the chances of fire occurring
- observance of advice from the state fire department and insurance fire surveyors
- special attention to high density storage areas and electronic equipment
- provision of suitable fire controls or warning devices, such as smoke detectors, heat detectors and alarm systems
- provision of firefighting devices, including sprinkler systems, hydrants and hose reels
- an evacuation plan
- fire drills at regular intervals
- training of personnel in the use of firefighting equipment and operation of an evacuation plan, and
- sources of information on technical aspects which affect fire safety at the workplace, such as the controlled use of flammable or combustible materials.

State regulations and ordinances set out many provisions on fire prevention and control, from the initial building design to requirements concerning explosives and dangerous goods. State regulations generally cover the provision of extinguishers and warning devices. Other regulations cover specific work processes, such as spray painting and the dipping of articles in flammable solutions.

What is fire?

Basically, fire is the result of the three factors required to support combustion, which are shown by the diagram below:

1. FUEL — solid, liquid or gas
2. OXYGEN — that exists in the air, and
3. HEAT — the ignition temperature will vary according to the material.

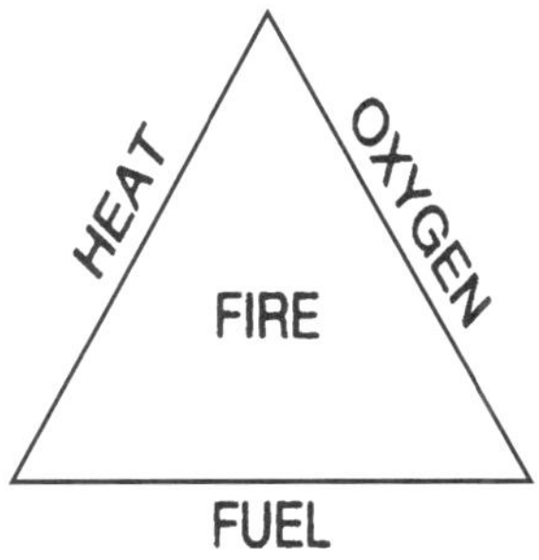

By removing one of these factors combustion cannot be supported (that is, by removing one side, the triangle will collapse).

By considering the triangle it can be seen that there are three methods of extinguishing a fire:

1. STARVATION — removal of the fuel or of combustible material in the vicinity of the fire to isolate it
2. SMOTHERING — reducing the oxygen content of the atmosphere in the immediate vicinity of the fire by excluding air, or
3. COOLING — lowering the temperature of the burning material below its ignition temperature.

The stages of fire

The development of a fire passes through four separate stages:

1. INCIPIENT — no visible smoke or flame or significant heat, but there is a condition generating combustion particles through chemical decomposition not visible to the human eye and develops over extended period
2. SMOULDERING — particles start to become visible as "smoke"
3. FLAMES — ignition point has occurred and infra-red energy is given off by flames, the visible smoke level decreases and more heat is developed, and
4. HEAT — this last stage develops very quickly from stage 3, with large amounts of heat, flame, smoke and toxic gases produced.

¶9-530 Causes of fire

There are a number of potential causes of fires at a workplace:

1. management failure to incorporate fire prevention and fire control as an integrated part of planning and procedures
2. actions by people, such as failure to extinguish a cigarette, or smoking in areas where flammable or combustible materials are present
3. conditions, particularly overloading electrical circuits and wiring, or incorrect fuse sizes
4. electrical malfunctions, such as in motors, switchboards or generators

5. unsuitable storage, such as combustible material near exhaust or steam pipes or high temperature areas
6. friction, such as that produced by conveyor belts or bearings
7. chemicals and liquids with low flash points that evaporate at room temperature to create an explosive vapour/air mixture, flammable gases and vapours, cleaning solvents and various dusts (flour/wood dust), and
8. unsuitable location of work processes and equipment.

Each of these aspects should come under scrutiny when the health and safety survey (see ¶4-150) is being carried out. Many causes on the above list may be regarded as "housekeeping" matters and can be included as part of the accident prevention program. All are preventable and can best be overcome by regular maintenance and inspection schedules of the equipment, materials, procedures and premises concerned.

Flammability of chemicals, liquids, gases and vapours should be checked out during the safety survey. Readily accessible information should be available through material safety data sheets and Hazchem Codes (see ¶10-240 and ¶10-260).

Regulations may provide for the provision of intrinsically safe extraction systems where explosive vapours and dusts result from the work process.

Hazards associated with high-piled storage

Results of American studies suggest that heat released from a 2.4-metre-high stack of burning combustibles was approximately 48,000 British Thermal Units per minute. If the height was doubled to 4.8 metres, however, the increase in heat multiplied nine times to around 440,000 British Thermal Units per minute. In tests of fires involving high-piled storage, roof temperatures have reached almost 900°C (steel loses strength and will have trouble supporting its load at less than half this temperature).

The other problem regarding high-piled storage is that many materials can collapse and so spread the fire to other areas or completely shield the fire from the sprinkler discharge. Depending on the weight of contents, it could of course help to collapse the surrounding walls or structure of the building in which the contents are contained.

The implication of these facts is that special attention should be paid to minimising areas of high storage of flammable materials and evaluating the degree of risks involved. State regulations or ordinances may also set storage restrictions.

Hazards associated with electronic equipment

The electronic components of computers, automatic storage systems, automatic processing plant and similar equipment are temperature sensitive, and their characteristics can be changed permanently if they are heated to temperatures beyond prescribed limits. These temperatures are lower than those that would render most other items of the plant inoperative. Smoke and corrosive products of combustion can impair the operation of electronic circuits to the point of major failure, necessitating complete replacement of equipment.

Electromechanical equipment can be damaged by minor fires that destroy cable insulation, damage relays and render machinery unworkable because components buckle under thermal expansion. Printed circuit boards can add fuel to a fire.

Any form in which electronic data processing information is stored can be distorted or melted by fire, or made faulty (to the extent that errors occur when used) by lesser heat.

The extent to which the organisation is dependent on the electronic equipment or stored data will indicate the capital outlay that is reasonable in order to provide protection against fire. Where the efficient operation of the organisation is dependent on the equipment and data, its destruction could prove disastrous.

Because of the great importance to an organisation of this type of equipment, it is advisable to seek specialist advice when purchasing and installing it, and considering the building layout in which it is to operate. Matters for consideration include wiring, paper storage, waste disposal, lighting, access, isolation from other risk areas, ventilation, detection and extinguishing equipment.

Some means of back-up record system should also be investigated, so that even if the equipment is destroyed, the organisation can continue to function.

¶9-540 Warning systems

Warning systems should be planned for in the original building design (or future alteration). If fire or any other alarm or warning devices are installed, employees should be aware of what the devices indicate and what they, the employees, are required to do. This could be done at the induction stage and reinforced by sign-posting on site and trial alerts. Regular testing of alarm systems should be carried out.

There are various types of alarms on the market, which can be operated either manually or automatically. Detection equipment is also available. Smoke/heat detection equipment can be incorporated into a system to act as a self-alarm or to notify the fire brigade.

Relevant government departments can advise on approved types of products, and equipment manufacturers can advise on prices.

¶9-550 Firefighting equipment

With fire extinguishers, it is important to understand the different types in use, and which particular type of fire each should be used for. The following table provides an indication of which type of extinguisher to use with the most commonly encountered types of fires.

Firefighting equipment

TYPE OF EXTINGUISHER	STANDARD COLOUR	TYPE OF FIRE				
		Class A	**Class B**	**Class E**	**Class F**	
		Wood, textiles, paper, rubbish	**Flam-mable liquids**	**Live electrical equip-ment**	**Cooking oils and fats**	**Comments**
Water	Red	Yes	No	No	No	Dangerous if used on electrical fires
Foam	Stainless steel with blue band Red with blue band	Yes	Yes	No	Yes	Dangerous if used on electrical fires
Wet chemical	Oatmeal Red with oatmeal band	Yes	No	No	Yes	Dangerous if used on electrical fires
Carbon dioxide	Red with black band	Yes	Yes	Yes	Yes	Not specially suitable outdoors
Powder	Red with white band (ABE)	Yes	Yes	Yes	No	Can cause serious damage to sensitive equipment
	Red with white band (BE)	No	Yes	Yes	Yes	
Vapourising liquid	Red with yellow band	Yes	Yes	Yes	No	Ventilate well after use

¶9-560 Overall responsibility for fire prevention and control

In each firm there should be one person with overall responsibility for all aspects of fire prevention from training to firefighting. Where this responsibility is divided among several people, the efficiency of the fire protection arrangements is diminished.

All organisations should therefore have a fire warden, who should be given full managerial backing. Where an organisation is small, a manager should be responsible for fire protection as well as his/her other duties.

One of the first jobs will be to draw up a fire training program for all employees, including temporary, part-time and seasonal. All the points covered in this guide should be included in the program, which must also take into account induction training for new employees.

The same person should have responsibility for handling other types of emergency, such as bomb scares.

¶9-570 Emergency evacuation plan

A plan for evacuation of the building in the event of an emergency should be formulated. The plan should be widely publicised (for example, on noticeboards and during each employee's induction program). Instructions should be clearly worded (in more than one language where appropriate) and include plans of the building layout showing locations of exits, meeting points and firefighting equipment.

The plan should be drawn up according to the organisation's needs and revised each time there are alterations to procedures, processes or premises. The following summary provides an indication of what items need to be included (see also ¶9-400).

Summary — emergency evacuation plan

1. Set out the escape plan, including:
   - (a) designated routes from each work location within the building, dividing staff numbers to use all exits and reduce congestion
   - (b) where to assemble once clear of the building
   - (c) whom to report to once clear of the building, and
   - (d) advice not to go back for valuables and not to waste undue time collecting valuables.
2. Prepare diagrams to set out high-risk areas such as kitchen, boiler/refrigeration/plant rooms, cleaners' cupboards, storage areas and hazardous processes.
3. Designate duties of individuals, such as checking corridors, using extinguishers, closing doors, and train accordingly. Ensure alternative staff are also identified in case someone is absent on the day.
4. Names and locations of emergency wardens should be listed on noticeboards.
5. Telephone numbers of fire station, police and ambulance should be listed on telephones.
6. On first discovering a fire, bomb, or threatening situation, notify emergency warden, switchboard, or fire brigade/police/ambulance (as required).
7. Encourage people to remain calm and orderly.
8. Feel surfaces of closed doors before opening them (to determine whether fire is present on the other side).
9. Close doors to fire escapes and windows behind you.
10. Carry out a "search and remove" operation floor by floor, including dressing rooms and toilets.
11. Shut down any potentially dangerous units, such as boilers.
12. Do not use lifts.
13. If escaping through a smoke-filled area, keep close to the floor (where there is most oxygen) and do not move too quickly.
14. If trapped, go to an outer room where the door can be shut and try to attract attention from a window.
15. Do not jump from a building unless there is no alternative, conditions make it safe to do so, or it is possible to land on a soft surface.

¶9-580 Fire drill

A well organised "disaster plan" will be of little use if employees are not aware of how they should react to a fire warning. Similarly, management will not know whether the plan will work effectively unless it has had a "test run". Even the latter, however, cannot simulate the fear/panic situation caused by hot smoke and fumes.

Employees need to be aware of the evacuation procedure (see ¶9-570), how to raise an alarm, and any designated role they are willing to accept. Examples of the latter include warden, part of a firefighting team, first aid attendant, or a particular duty such as closing a door or checking a corridor. It is advisable to medically clear each volunteer to ensure that they have no physical impairment, such as heart or respiratory problems.

There is no set period at which a fire/emergency drill should be held. It may vary according to labour turnover rates and organisation size. It is important, however, that management be confident that employees are sufficiently well aware of the procedure so that any emergency will be handled efficiently.

¶9-590 Fire prevention and control checklist

HAVE YOU CHECKED ... ?

1. *Housekeeping* — waste containers, combustible material containers, storage, handling of flammable liquids and substances, rubbish removal
2. *Smoking* — enforcement of rules, "no smoking" signs
3. *Electrical equipment* — wiring, motors, fuses, fuse panels, switch boxes, cleaning
4. *Fire doors and exits* — door condition, lack of obstructions, kept open, closing devices, sign-posting, fire drills
5. *Fire extinguishers and hoses* — pressurising, charging, accessibility, servicing, employee training
6. *Sprinkler systems* — control valves, lack of obstruction, anti-freeze provisions, alarms, inspection of sprinkler heads
7. *Hydrants* — accessibility, condition, drainage
8. *Water supplies* — valves, connection, pump condition, tanks filled, gravity tanks
9. *Automatic alarms and extinguishers* — instructions to employees, condition of equipment
10. *Evacuation procedures* — practised and known by all employees

The above list is not intended to be exhaustive but provides an indication of the matters that should receive attention when checking for safety against fire.

Further references

AS 4655-2005: *Fire safety audits*, Standards Australia, 2005.

CS FP 001-1995: *Fire Emergency Response*, Standards Australia.

Fire Safety: an Emergency Guide, Health and Safety Executive, London, HMSO, 1999.

Davletshina, TA, *Industrial Fire Safety Guidebook*, Westwood, New Jersey, Noyes Publications, 1998.

Fire Engineering Guidelines, Fire Code Reform Centre Ltd, Sydney, 1996.

Fire Safety in Tall Buildings, Council on Tall Buildings and Urban Habitat, Tall Buildings and Urban Environment Series, Bethlehem, Pennsylvania, McGraw-Hill, Inc, 1992.

Fire Safety Journal, International Association of Fire Safety Science, Oxford, Elsevier (the official journal of the International Association of Fire Safety Science).

Paterson, T, *Offshore Fire Safety*, Tulsa, Oklahoma, PennWell Publishing Company, 1993.

Further information on fire safety can be obtained from the fire brigade in each state, or from the Fire Protection Association Australia.

ALCOHOL AND OTHER DRUGS

¶9-620 Extent of the problem

Alcohol addiction is ranked as the fourth most prevalent disease in Australia after cancer, heart disease and mental illness. Alcoholics may be found at all levels of the workforce. The International Labour Office has reported that 15–30% of all work fatalities are related to alcohol or other drugs, and employees with alcohol or drug problems have 200–300% more absenteeism than other employees.

An alcoholic may be defined as a person whose physical and emotional health, job and general responsibilities towards society are repeatedly adversely affected by the continual abuse of alcoholic beverages.

"Occupational alcoholism" relates to recurrent poor job performance resulting from continued excessive drinking.

The problem appears to be most widespread in the age group 30–50.

Effects on work

Three main effects are characteristic of problem alcoholism amongst employees:

1. *Absenteeism.* This grows more pronounced as the problem deepens. Initially the problem drinker may try to avoid taking time off, but later takes off more and more time including "off-the-job" absenteeism (prolonged lunch hours, etc) and gives feeble excuses for absences.
2. *Accidents.* There is an increase in mistakes and misjudgments which can lead to accidents.

3. *Job performance*. Initially job performance does not suffer — it may even be above the average, as the employee tries very hard to do well in order to conceal the problem. However, if the alcohol dependence deepens, performance will fall off sharply.

Exposure to some kinds of hazardous substances at work (such as some solvent vapours) can magnify the effects of alcohol and result in a much higher level of impairment than would be expected. These types of combined effects can be very dangerous, particularly if people are not aware of the potential interaction.

Performance can also be impaired by other drugs, both prescribed and illicit. Tranquillisers used to counteract anxiety and insomnia can adversely affect motor coordination, judgment, reaction time and mental processes — especially if combined with alcohol.

Some analgesics, for example codeine, can affect the ability to use machinery safely. Some stimulants, such as amphetamines (known as "speed"), can impair performance even while increasing alertness and delaying fatigue. Marijuana (cannabis) can affect motor coordination, sensory function, perception, short-term memory and tracking ability. As a result, it can have significant effects on the safe job performance of people driving vehicles or operating potentially dangerous equipment.

¶9-630 Detecting the problem

There are various physical, psychological and social signs which, together with poor work performance, may help to identify someone with a drug/alcohol problem. Signs of problem drinking can include:

- absenteeism
- more unusual excuses for absences
- lower quality of work
- mood changes after lunch
- red or bleary eyes, or flushed face
- hangover on the job
- irritability
- avoiding boss or associates
- smell of alcohol on the breath.

The emphasis of a successful rehabilitation program lies in the early discovery of clues pointing to an alcohol problem. This is when the alcohol-troubled worker will have the best chance of recovery.

In addition to those listed, other signs of a possible alcoholic include forgetfulness and memory blackouts, dishonesty, temper outbursts, intolerance and hypersensitivity about matters related to drinking, and suspicion of others, including management.

Of course, many of these signs can be symptomatic of non-alcohol-related problems as well. Referral of an employee to an employment assistance program (EAP) due to poor work performance can result in these problems being handled appropriately. The philosophy of such programs is to consider all problems that are affecting an employee's performance.

¶9-640 Role of the employer

On a cost/benefit basis, various studies indicate that treating the problem is much more successful than dismissing and replacing the employee.

A risk management approach should be adopted, ranking risks in terms of their health and safety consequences. It is important to distinguish between cases where there is the potential for serious accidents, death or injury, and cases where the employee's efficiency may be reduced but safety consequences are not really an issue. In the latter type of case, if there is a performance problem it should be dealt with as such. In the former type of case, however, where an employee's drug or alcohol problem represents a risk to their own safety or that of others (including employees and others present at the workplace), employers are bound by their duty of care to control the risks.

In general, the most effective method of treating employees with alcohol and drug problems is to establish some sort of EAP. Some organisations use specific alcohol and drug programs, but EAPs have the advantage of taking a broad approach and dealing with any personal problems a worker may have. For this reason, EAPs have become the preferred model of intervention into alcohol and drug problems at the workplace.

It is necessary to set up procedures for identifying and dealing with cases of alcohol- or drug-abusing employees and, for reasons of consistency and fairness, ensure that these are observed.

The role of the employer is to identify the problem and encourage and arrange access to treatment. The actual provision of counselling services should be left to experts. An organisation may wish to establish internal counsellors, or it may use outside resources (see ¶9-660). Part of the employer's role will be to ensure that supervisors have sufficient training to be able to identify employees in need of treatment. The chart at ¶9-630 provides some guide to the signs supervisors should watch for.

¶9-650 Employee assistance programs

Employee assistance programs (EAPs) are aimed at providing assistance to employees with any one of a wide range of personal problems, including drug and alcohol problems. Such programs are based on the premise that employees cannot and do not leave their problems at the front door of the workplace and then pick them up again on the way home. Personal problems of employees can affect their performance at work.

Personal problems that can be dealt with by an EAP include:

- drug and alcohol problems
- marital and family problems
- financial worries

- gambling
- interpersonal/social problems
- health, and
- stress/lack of work satisfaction.

The key features of an EAP are:

- policy
- procedures for referral, and
- counselling.

The employer may wish to publicise the availability of the program, and to establish a mechanism to evaluate its success.

EAP policy

The aim of any EAP policy is to set out the goals of the program, and to alleviate fears about the methods used. The confidential nature of any referral and counselling needs to be stressed.

The following features should be part of any EAP policy:

1. The aim of the EAP is to help employees restore their health and work performance to satisfactory levels.
2. The EAP is based on the philosophy that the employer is able to expect a satisfactory work performance, not that the worker's personal problems are the business of the employer.
3. No penalties of any sort will apply to any employee seeking assistance through an EAP.
4. Employees are encouraged to voluntarily refer themselves for assistance.
5. Supervisors should, as a matter of course, offer the EAP to any employee to whom they speak about poor work performance.
6. All details relating to counselling and treatment are to be treated with strict confidentiality.
7. Employees requiring treatment for a personal problem may use any accrued sick leave entitlements. Where no sick leave or annual leave is available, the employee will be entitled to leave without pay.

In addition, the policy may set out the range of problems dealt with by the EAP (see the list above).

Procedures for referral

Referral to an EAP is the key step in ensuring the success of the program. While supervisor-initiated referral may be the main mechanism of access, scope should always be made for self-referral to the EAP.

The following steps could be used by supervisors:

1. When interviewing an employee about work performance problems a supervisor should suggest that the employee is able to visit the EAP counsellor if there are personal problems that the employee does not

wish to discuss with the supervisor. The decision whether to see the EAP counsellor remains with the employee.

2. If interviewing an employee for a second or third time, the supervisor should again offer the EAP to the employee, with a reminder that there could be disciplinary action if the work performance problems persist.
3. At the point of administering disciplinary action, the supervisor may either
   (a) offer the EAP in lieu of disciplinary action, or
   (b) initiate disciplinary action and offer the EAP.

Some supervisors may feel reluctant to suggest the use of an EAP, as they feel the suggestion itself is a form of judgment. This can be overcome by stressing that the suggestion is based on a decline in work performance and is in no way a moral judgment.

Where the employee elects to attend counselling during working hours, the supervisor is entitled to know that the employee attended the counselling (and no more). If an employee refers himself/herself to the EAP and elects to attend counselling outside of work hours, there is no need for the supervisor to be informed.

¶9-660 Who is involved in counselling?

Counselling is at the heart of an EAP. Counsellors may be either specialist employees of the organisation, or belong to external agencies.

Counselling is aimed at helping individuals to identify strengths and problem areas in their lifestyles, and to make decisions about positive steps to resolve those problems.

Employee assistance program counselling is usually limited to one or two sessions. It is not designed to be long-term therapy. If further assistance is necessary, the counsellor may refer the employee to a specialist. Any further assistance will then be at the employee's expense.

Internal counsellors

Most commonly, counsellors are qualified psychologists with a clinical background or social workers with extensive counselling experience.

Counsellors need to have a background in a wide variety of personal problems, including drug- and alcohol-related problems. Other staff within the organisation, such as the human resources officer or medical staff, are not usually appropriate.

Internal counsellors must be perceived as having a high level of independence and credibility within the organisation.

Most organisations with fewer than 2,000 employees cannot justify the employment of a full-time EAP counsellor. In this case, part-time counsellors may be employed, or the organisation may contract with an external agency.

External counsellors

When contracting with external counselling resources, it is usual for an organisation to pay an annual retainer rather than a fee for each individual. The latter option would compromise an employee's confidentiality if the employee chose to use the counsellor outside of work hours.

External counsellors may suffer from a lack of familiarity and contact with the organisation they are serving, if sufficient orientation and access is not provided when establishing the relationship.

Counselling and disciplinary procedures

In safety-critical occupations such as pilots or bus drivers, it is not enough to leave the resolution of drug and alcohol problems to voluntary, long-term processes. In this situation, the organisation will have to deal with it in a more active manner.

The approach will vary according to the extent of the problem, the likely consequences, the position of the employee and the nature of the industry. Counselling and disciplinary procedures should be set out in the organisation's policy. They should be consistent with any existing awards, agreements or procedures which apply in the workplace, and they should be communicated to employees.

A staged series of interviews can be used to help steer the employee away from irresponsible or unsafe drug or alcohol use. In safety-critical jobs, the possibility of termination of employment may be considered if less drastic measures do not result in an acceptable level of risk control.

¶9-670 Drug and alcohol testing

Drug and alcohol testing in industry is a sensitive matter, and issues of the rationale and validity of such tests must be carefully addressed. In Australia, testing has been considered almost exclusively in the context of safety rather than impairment of efficiency. It is generally held in Australia that there are only particular situations where the employer has the right or the responsibility to be concerned with the use of drugs (including alcohol) during working hours.

It would seem to be difficult to justify testing for drugs in relation to impaired work performance when work performance can be directly observed and can otherwise be monitored through existing accepted procedures. The issue becomes one of the validity of drug testing in terms of indicating or predicting unsafe behaviour.

Problems with testing

There are technical problems in drug testing relating to the reliability of tests (through both false positive and false negative results). In addition, for some types of drugs a positive test result does not necessarily relate to current impairment. For example, the metabolite of marijuana can be identified and produce a positive test result several weeks after the effect of the drug on functioning has worn off. A negative result can also be unreliable as an indicator of the ability to function safely — for example, work performance can be markedly impaired 14 hours after a drinking session, when no blood alcohol can be detected.

From the point of view of both employers and unions, what is needed is a means of reliably and validly assessing employees' abilities to function (perform) safely, without intruding into their otherwise irrelevant personal habits. That is, it is the employees' functioning that is of direct concern rather than their bodily fluids. A focus on function is the most valid way to approach the issue, because factors other than alcohol or other drugs may affect an employee's concentration or motor coordination, for example, fatigue, personal stress or illness.

Therefore, when testing in principle is agreed to, it is more valid, reliable and useful to test **functioning** rather than testing for drug levels. Simulation tests have many advantages, both in principle and technically, over alcohol and drug testing.

Due to its questionable validity, and problems in reliability, privacy and logistics, drug and alcohol testing by employers has understandably become an industrial issue in Australia. Testing is currently opposed by the Australian Council of Trade Unions (ACTU) on various grounds, including the argument that priority should be given to removing potential causes of such problems in the organisation and to the design of safe work systems and environments.

Further references

AS/NZS 4308:2001: *Procedures for the collection, detection and quantitation of drugs of abuse in urine*, Standards Australia.

Allsop, S, Phillips, M and Calogero C (eds), *Drugs and Work*, Melbourne, IP Communications, 2001.

Nicholas, R, Allsop, S, Calogero, C, Phillips, M and Ormonde, N, *Alcohol and Drugs in the Workplace; Issues, Trends and Practices*, Perth, Chamber of Mines and Energy of Western Australia, 1996.

Roman, PM and Blum, TC, "The workplace and alcohol problem prevention", *Alcohol Research & Health*, 2002, 26(1).

Guidelines for Drugs, Alcohol and the Workplace, WorkCover Corporation of South Australia, 2001.

Guidance Note: Alcohol and Other Drugs at the Workplace, WorkSafe Western Australia Commission (undated).

Buon, T and Compton, B, "The development of alcohol and other drug programs in the workplace", *J Occup Health Safety — Aust NZ*, 1990, 6(4): 265-267.

National Guidelines for Employee Assistance Program, National EAP Executive, Canberra, 1989.

Drugs, Alcohol and the Workplace, WorkCover Authority of NSW.

Not at Work, Mate (video and resource kit), National Occupational Health and Safety Commission, 1994.

Alcohol and Drugs in the Workplace, Queensland Government, Workplace Health and Safety, 2001. (Website at www.whs.qld.gov.au)

See also the websites of government OHS authorities and other OHS information providers such as those listed in the Appendices.

FITNESS FOR WORK

¶9-700 What is fitness for work?

A workforce able to competently fulfil its responsibilities is one of the prerequisites for safe, healthy and productive work, along with safe systems of work and suitable equipment.

Fitness for work may be thought of in terms of physical fitness for the physical (and especially manual handling) demands of the job. In its broader sense, however, it also includes a range of matters such as fatigue/hours of work issues, impairment of work performance by alcohol or other drugs, as well as recruitment/selection, recovery from injury, training, provision of information and supervision.

Fitness for work is not just an issue associated with individual employees. Systems of work and equipment both contribute significantly to fitness for work, for example through shift arrangements and ease of use of equipment. This section therefore addresses how these and other factors which may reduce employee fitness for work can be managed to best advantage for the organisation and its staff.

Health and safety risks from lack of fitness for work

Employees who are fit for work are in a physical and mental state that will allow them to fulfil their work responsibilities competently and in a manner which does not compromise safety. On the other hand, workers whose work performance is impaired by fatigue, by drugs (including alcohol and prescribed medication), by stress or by other factors may be more likely to put themselves or others at risk.

The importance of fatigue was highlighted by an April 2003 study conducted by the US Army Research Institute of Environmental Medicine. It found that tired soldiers' reactions were worse than if they had been drunk or sedated, with slower reaction times, reduced vigilance and problems remembering details. In a quick-decision test, fatigue resulted in an increase in errors, from one or two to more than 15.

Many other aspects of the workplace can also have a significant influence on fitness for work. Employees who are not fatigued and who have not used drugs or alcohol will still not be able to fulfil their work responsibilities competently if they have not received adequate training or if the equipment they are expected to use is not fit for the purpose. Similarly, the design of the job may be such that most workers will be too tired by the end of the shift to be fit to meet the physical challenges they are confronted with. This means that workplaces should take a holistic approach to fitness for work, examining how they manage all of the influences on fitness for work, rather than focusing entirely on individual factors.

A range of work-related risks arise from poor fitness for work, whatever the source.

Increased risk of accidents/incidents

Fatigue causes impairment of cognitive and physical functions, leading to increased potential for incidents due to human error. Sleep deficit, for example, can cause significant lapses in concentration, including episodes of involuntary "napping" (micro-sleeps). Such episodes can occur at any stage of a shift cycle with obvious negative consequences for safety, though they are more likely in some shifts than others. It has been asserted that working at night, eg between the hours of 4 am and 8 am on the first night of night shift, can be the equivalent of having a blood alcohol reading equal to or greater than 0.05.

As this suggests, the impairments caused by fatigue and drug/alcohol consumption can be the same. Concentration lapses are a common result of alcohol and drug consumption too. Even where concentration does not lapse, impairments caused by drug/alcohol consumption can result in slower response times because of reduced physiological arousal. It can also result in reduced cognitive function and capacity to make decisions and loss of situational awareness (not being able to see the "big picture" or to recognise that one is impaired). This means that at the same time that errors are more likely, there is reduced margin for errors. Workers who are not fit for work because of fatigue or drug and alcohol consumption may not be able to pay sufficient attention to secondary tasks, eg they may operate a forklift truck correctly, but drive too quickly. They may be unable to recognise the existence of a problem or may only recognise it too late to do anything about it (see **Alcohol and other drugs**, ¶9-620–¶9-670).

Fatigue, as well as alcohol and drug consumption, can significantly impair hand–eye coordination so that even if affected workers are able to identify the need to respond, they may not be able to respond correctly.

All of these problems are more likely where work is repetitive, tedious or requires sustained vigilance, as in equipment operation. Very low levels of physical activity can increase workers' sleepiness, while at the other end of the spectrum jobs that are very demanding from a manual handling point of view can also increase the risk by increasing workers' fatigue. For example, it has been shown that nurses are more prone to lifting injuries in the last quarter of their shift.

Musculoskeletal problems

Lack of physical fitness for work can increase the risk of musculoskeletal disorders such as back injuries, strains/sprains and injuries to neck, shoulders, knees, etc. Problems of this type often constitute a high proportion of work injuries, and a major cost to the organisation.

Poor work performance

Poorer work performance can also result from lack of fitness for work. For example, train drivers adversely affected by shiftwork are more expensive drivers; they use more fuel, brake more frequently and their driving techniques generate greater "in train" forces. While not strictly speaking an OHS issue, work performance and quality issues can also feature amongst the benefits for employers when fitness for work is addressed.

Factors which may affect fitness for work

Many factors may affect fitness for work, and these can often interact with each other:

- physical fitness
- fatigue — due to shifts, working hours, nature of work, activities outside of paid work
- consumption of alcohol or other drugs (prescription and non-prescription, legal and illegal)
- psychosocial factors (eg psychiatric problems, unresolved domestic problems)
- hazardous substances or environments
- recruitment/selection
- training, provision of information and supervision, and
- previous injury.

These factors also interact with aspects of the work environment such as ease of use of equipment, noise levels, ventilation or the amount of physical activity built into the job.

Physical fitness

Aspects of physical capacity such as muscle strength, aerobic capacity (stamina) and mobility/flexibility of joints can be influenced by many factors, such as youth, age, pregnancy, return to work after a period away, or after an injury, and lack of warm-up exercises before physically demanding work.

Fatigue — due to shifts, working hours, nature of work, activities outside of paid work

Fatigue can be defined as increasing difficulty in performing physical or mental activities. Fatigue results from insufficient rest and sleep between activities. Work-related fatigue can arise from non-traditional shifts, greater workload within standard shifts or from situations requiring concentration for extended periods during work hours, working in temperature extremes or working with high-risk situations. Causes of non-work-related fatigue include sleep disruption due to ill family members, stress associated with financial difficulties or domestic responsibilities. Moonlighting (working at a second job) can significantly increase fatigue, as can long periods of driving to and from work.

Work-related fatigue can be reduced by good management of the organisation. Non-work-related fatigue varies considerably between individuals because of all the different circumstances, and is dependent on a person's environment as well as physical and mental attributes. Non-work-related fatigue is best managed at the individual level through strategies such as education and counselling, as well as identifying any organisational issues which might have an impact.

A common cause of fatigue in shiftworkers (see **Shiftwork** in Chapter 11) is the disruption caused to the body clock (''circadian rhythms'') by only having the opportunity to sleep at times when the body is not used to sleeping. This body clock never completely adjusts, even in people who work night shift permanently. As a result, getting adequate undisturbed sleep during the day may be difficult during periods of night work. Extended hours that carry into the night time may create a similar problem. The result is sleep deprivation, which may lead to what is called sleep debt. If workers are seriously sleep deprived, they may suffer micro-sleeps. A micro-sleep is a brief involuntary ''nap'' that lasts for about four to five seconds.

Symptoms of fatigue include:

- tiredness
- loss of alertness
- greater likelihood of involuntary sleeping (eg micro-sleeps)
- poor memory (eg not being able to recall the work activity which has just been completed)
- irritability and depression
- poor judgment, and
- slow reaction to signals.

Acute fatigue results from an immediate episode of sleep deprivation. It results in:

- unpleasant muscular weariness
- tiredness in everyday activities, and
- reduced coordination and alertness.

If the sleep deprivation is not remedied, chronic fatigue can result which leads to further deterioration in performance.

Hours of work require managing so that long-term health problems do not develop and so that adverse consequences for fitness for work do not result.

Consumption of alcohol or other drugs

Because they impair decision-making and performance, consumption of alcohol and prescription and non-prescription drugs can adversely affect fitness for work. Use of alcohol and drugs is not always independent of work.

Work-related stress as well as more benign causes such as social events with workmates can result in consumption patterns that cause impairment. For these reasons, employers should address this issue on an organisational as well as an individual level (see **Alcohol and other drugs**, ¶9-620–¶9-670).

The effect of drug consumption on fitness for work relates to prescription medications too. For example, workers' safety in driving or operating machinery may be compromised if they have taken medication which results in drowsiness. In addition, prescription guidelines generally assume that the medication will be used during the day and that sleep occurs at night. For shiftworkers this may not be the case. The effects of medications can be related to the time of day at which they are taken. Therefore, depending on the individual roster, the intended effect of the prescribed medication may vary from the actual effect. Consumption of alcohol or other drugs may interact with the effects of prescribed medications, and also with exposure to some workplace hazardous substances, further increasing the risk of errors or accidents.

Workers should advise medical practitioners about their working hours and type of work when being prescribed medication.

Psychosocial factors

The psychological and social situation of individuals affects fitness for work when it results in stress and fatigue. As discussed above, fatigue can result from family and social situations placing undue demand on individuals. This might include carer responsibilities and family dislocation as a result of shiftwork. Employers need to address these issues from an organisational as well as an individual perspective. Employees should not be discriminated against because of their status as a carer.

In many cases, shiftworkers find it difficult to maintain a satisfactory social and family life and many neglect recovery from work in order to fulfil other needs. The difficulties created for family and social life may result in pressures on relationships, on domestic workloads and on community activities. These problems can also exist with day workers.

Psychosocial stress can result from:

- work stress causing further pressures at home
- body clock disruption from shiftwork
- disruption of social and family life as a result of shiftwork or other factors
- sleep disruption, and
- irritation and strain from work and family conflicts.

Work scheduling that is based solely on organisational requirements may not acknowledge the personal, domestic and social needs of the workforce.

Different causes of poor fitness for work can reinforce each other. For example, fatigue can lead to greater misuse of alcohol and drugs. Equally, negative social conditions can exacerbate fatigue.

Stress

Stress from both organisational and personal causes can result in the kinds of impairment listed above as resulting from fatigue and can also lead to depression and increased use of alcohol and drugs (prescription and non-prescription) which damage fitness for work even more.

Hazardous substances and environments

Exposure standards for hazardous substances and environments (eg noise) have been determined on the basis of eight-hour working days (not nights) and 40-hour working weeks. As is the case with prescription drugs, exposure at a time when the body's circadian rhythms are set for sleep may have quite different and perhaps unexpected effects than exposure of the same quantity and duration during the day. Furthermore, exposure of greater duration, say 12 hours whether at night or in the day, can have an increased effect over the traditional eight-hour shift that is not simply a 50% increase.

Recruitment/selection practices

The recruitment and selection process aims to choose the most suitable candidate for the job. If the requirements of the job are not defined realistically and carefully, or if the selection process does not fulfil its aims, a job–person mismatch can occur, which can be costly to both the individual and the organisation.

Training, provision of information and supervision

Even the most suitable employee, equal to the demands of the job, and unimpaired by fatigue or other factors, can fail to competently fulfil work responsibilities if he or she has not been provided with adequate training and information or if supervision is not adequate.

Previous injury

Injured workers who have no major pathology or injuries requiring surgical intervention may be limited in terms of their physical capacity to carry out the duties of their position.

¶9-720 Key requirements for fitness for work

The employer's duty of care requires a proactive approach to workplace health and safety involving attention to all matters which may increase risks, including lack of fitness for work. Management needs to make appropriate provision for addressing the impacts of:

- work-related fatigue
- drug/alcohol consumption
- job design and the work environment, and
- recruitment/selection and other personnel practices.

Where fitness for work issues interface with other human resources matters such as pregnancy or special job requirements, anti-discrimination considerations must also be addressed.

In some areas there are specific legislative requirements relating to fitness for work. For example, in Western Australia, the Occupational Safety and Health Regulations 1996 were amended from 1 July 2003 with new requirements relating to drivers of commercial vehicles.

The new provisions require the development of a driver fatigue management plan, and a medical practitioner must examine and pass commercial vehicle drivers as fit according to the *Medical Examinations of Commercial Vehicle Drivers*, published by the Federal Office of Road Safety and the National Road Transport Commission, or equivalent requirements.

In Queensland, Workplace Health and Safety issued a *Fatigue Management Guide* in February 2005. This guide was developed to raise awareness of fatigue and its effect on work and health, noting specifically shiftwork conditions and extended working hours.

New South Wales has also proposed new requirements in the draft Occupational Health and Safety Amendment (Long Distance Truck Driver Fatigue) Regulation 2005. The regulation would impose obligations on employers and persons who contract with self-employed carriers to assess the risk of harm from fatigue, and eliminate or control the risks.

¶9-730 Principles for managing fitness for work

Risk management

The employer holds the fundamental responsibility for managing the risks associated with fitness for work. However, employees are responsible for ensuring that their behaviour does not create or exacerbate risks. In particular, the parties have the following responsibilities:

	Employer	Employee
Identify	• To identify the risks associated with fitness for work, including establishing and monitoring an ongoing system for identifying problems.	• To report problems with fitness for work. • To contribute to establishing and monitoring reporting systems.
Assess	• To assess the risks associated with fitness for work.	• To contribute to the assessment of risks.
Control	• To control the risks associated with fitness for work, specifically: — ensure that work systems are flexible enough to deal with factors influencing fitness for work (eg allowing supervisors to reschedule tasks if fatigue becomes a problem)	• To cooperate with control measures.

	— ensure that shift arrangements control risks, especially fatigue — ensure that work arrangements do not contribute to alcohol and drug consumption or interfere unreasonably with employees' capacity to meet family and social commitments — provide training, information and supervision which support effective management of fitness for work — ensure that fitness for work testing does not shift responsibility for managing the risks to the individual.	• To contribute to the design and implementation of control measures. • To manage individual factors which affect fitness for work (eg ensuring adequate rest between shifts, controlling alcohol and drug use).

The fact that individual behaviour outside of work can have a considerable influence on fitness for work does not reduce the employer's obligation to address the issue consistently with the principles of risk management, including application of the hierarchy of control.

The three steps of risk management (identify, assess and control) are essential to an effective approach to dealing with fitness for work. The sensitive nature of many of the issues involved in this area mean that how the risk management process is undertaken is at least as important as what is done.

Experience suggests that the rigid application of rules is counterproductive and that the approach must be flexible enough to respond to the views of the workforce. Sensitivity to the communication and consultation needs of those who work shiftwork is also critical — those who work nights and weekends may require special attention in communication strategies. Risk management must be implemented in a consultative, participative manner, involving all affected workers throughout the process and in decision-making about the outcomes. Given the nature of the issues being dealt with, imposing risk control measures without consultation is unlikely to result in effective control of risks.

Risk identification

The following risk factors should be considered when identifying the risks associated with fitness for work.

Workload

All jobs, and all tasks within jobs, impose a workload. Factors which should be taken into account in designing jobs, and in identifying risks to fitness for work, include:

- the physical effort required, eg lifting, carrying, driving, operating a machine, using a keyboard
- the demands caused by monotonous, repetitive activities or the need for high vigilance and concentration

- the pressure on the employee, eg number of different tasks required to be done where realistically only a certain number can be completed effectively, the amount of control and autonomy the employee has over the work
- the hours of work including night shift versus day shift, length of shifts, timing of shifts, and breaks between and within shifts
- the body's physiological and psychological responses to workload, eg heart rate, hormone production, blood pressure, stress, anxiety and fatigue
- whether job performance changes with the time of day
- whether supervisors and workers are able to change work activities and tasks in response to fatigue, and
- the work environment, eg heat, cold, dust, noise.

Moonlighting (working at other jobs) can add significantly to an individual's workload. It can be difficult to determine but the consequences it can have for fitness for work means that the possibility of moonlighting should be identified. Use your consultative forums to address this issue, in particular to develop an agreed approach.

Travel time

In addition to the actual contract hours at work, travel time to and from work each day should be considered, particularly when extended shifts are worked. In some instances, travelling time can add significantly to a work shift. Sleep time is reduced and recovery times needed after work may be extended. Hazards associated with driving on long straight stretches of road are a problem, especially when fatigued.

Particular risks of incidents

Situations where fitness for work is critical should be identified to highlight where workers need to perform at their optimum and where fatigue and other impairments may play an important role. These will include situations where:

- work error is dangerous to society
- work error is dangerous to the worker or workers themselves
- a failure to respond is dangerous to society, and
- failure to respond is dangerous to the worker or workers themselves.

Family and social needs of employees

The family and social commitments of employees should be taken into account in risk identification. This does not mean that employees with family and other commitments should be unfairly targeted. It is illegal to discriminate against employees on the basis of their status as a carer or their religious beliefs. Arrangements for shiftwork and extended hours should consider the needs, commitments and responsibilities for all employees who are trying to balance work and other commitments.

Alcohol and drug use

The preceding risk factors can also lead to increased risk of hazardous patterns of alcohol and drug use, eg short breaks between shifts may mean that prescription drugs are more likely to be used to encourage sleep. The presence of other risk factors should also be identified. These include the likelihood of alcohol consumption associated with work (eg the presence of a bar, or work-related social events where alcohol is offered). Often, workforce representatives will be able to identify whether alcohol and drug use is a risk. Again, consultative forums should develop an agreed approach to this issue (see **Alcohol and other drugs**, ¶9-620–¶9-670).

Risk identification relating to alcohol and drug use does not require that individuals who have problems with alcohol or drug use should be unfairly targeted. Indeed, any approach that seeks to punish these employees is unlikely to control the problem effectively. This means that random breath testing and other drug tests, if they are used at all, need to be used as part of an agreed and holistic approach to managing these risks. Testing may be useful as a tool within a thorough risk identification process, but should not be relied upon as the only risk identification technique. Testing whether someone has used drugs will not determine whether the person is fit for work or not. It will only reveal the use of drugs at some stage prior to the test, which may have no relevance to work performance.

Assessing risks

Risk assessment should involve consideration of the likelihood of serious consequences from the existing approach to dealing with fitness for work. As well as the data generated by identifying the risks, information that will support effective risk assessment includes:

- incident reports
- self-reports from employees
- aggregate data from any EAPs (ensure that confidentiality is maintained in using such data), and
- advice from those with technical expertise in the relevant disciplines.

Controlling risks using the hierarchy of control

Controlling the risk factors identified and assessed in the previous two steps should be in accordance with the hierarchy of control. Controlling organisational factors which can adversely affect fitness for work is best achieved with higher order controls (eliminate, substitute and engineering). However, administrative arrangements such as job design, training, supervision and recruitment/selection practices can also have a significant impact on fitness for work.

Non-organisational or individual factors which can adversely affect fitness for work can also be addressed via administrative arrangements such as EAPs.

Eliminate

Eliminating risk factors associated with poor fitness for work can be difficult — it is not always possible or even desirable to eliminate shiftwork and excluding workers because of their family and social commitments can be counterproductive and illegal. However, wherever possible, the sources of high risks should be removed from the workplace. This might include:

- eliminating the use of shiftwork or extended hours for certain functions
- moving as much activity as possible to day shifts, particularly work which may be a high risk at times when impairment is likely (eg at night)
- removing any sources of alcohol from the workplace, or
- eliminating sources of risks that might exacerbate poor fitness for work (eg lack of job control, excessive noise or manual handling, extremes of temperature or poor ventilation).

Job design should ideally ensure a sufficient level of physical activity or variation of types of activities to help workers maintain alertness, thus eliminating to some extent the likelihood of lapses in concentration, micro-sleeps, etc.

Substitute

Substituting safer practices for those currently used can produce an effective set of control strategies. Such strategies include:

- improved work scheduling
- improved roster design, and
- control over hours of work.

Engineering

Engineering controls may help control risks, for example:

- heating and cooling to control ambient temperatures and suitable ventilation to support alertness
- alarms and monitors, particularly for solo work (eg driving vehicles), and
- ergonomic modifications of equipment or workplace layout. For example, if machinery, benches or workplace layout were designed for taller people, they may be more tiring for people of small stature to work with. A modification of the equipment or layout (such as a platform to stand on) can reduce the worker's fatigue, thereby improving fitness for work.

Administrative systems and work practices

Effective selection, recruitment and placement policies and practices can improve the job–person match without discriminating unfairly. By establishing suitable training programs and providing suitable information and supervision, adherence to safe work practices can be consolidated. Health promotion programs, including opportunities to help workers improve physical fitness or quit smoking, can also contribute to a fitter workforce.

An effective EAP is generally recommended as a key strategy for dealing with individual causes of poor fitness for work, including drug/alcohol consumption, personal and psychosocial matters. Essential features of effective EAPs are:

- a "no-blame" approach
- application to all employees, including management
- protection of individual rights, including employment security and privacy
- clear guidance about how referrals to the EAP will occur (eg whether only voluntary or will individuals with particular problems be required to seek help), and
- agreed procedures for dealing with non-compliance with treatment regimes.

See ¶9-650 for more details on EAPs.

Training

Information and education for workers and management are critical supports for any control strategies. Employees should be provided with information on the effects of drugs and alcohol and on the agreed approach to this issue, as well as other relevant issues such as fatigue management.

Provision of training and advice should include managers and supervisors who should be fully educated about OHS issues of fitness for work. Supervisor training should cover responsibilities to monitor workers' fatigue levels and check for other signs of impaired performance, especially in safety-critical jobs.

Training on the workplace's agreed approach to managing fitness for work is a critical part of an effective strategy.

Protective gear

The use of personal protective equipment is of limited use as a control measure in relation to fitness for work. It may help in some circumstances, however, eg the use of hats and sunglasses can reduce environmental stress and fatigue in outdoor workers.

Further references

Fatigue Management Guide, Workplace Health and Safety, Queensland, February 2005.

"New standards for truckies designed to save lives", *Safetyline* 56, October 2003, WorkSafe Western Australia. (Website at www.safetyline.wa.gov.au/pagebin/wswaslmg0374.pdf)

Fatigue Management for Commercial Vehicle Drivers, Code of Practice, WorkSafe Western Australia, 1998.

Donoghue, AM, *The Calculation of Accident Risks in Fitness for Work Assessments: Diseases that Can Cause Sudden Incapacity*, Minerals Industry Safety and Health Centre, University of Queensland, October 2000.

Dawson, D, Fletcher, A and Hussey, F, *Burning the Midnight Oil: Parliamentary Inquiry Into Managing Fatigue in Transport*, Centre for Sleep Research, University of SA. (Available at www.ir.qld.gov.au)

See also the websites of government OHS authorities and other OHS information providers such as those listed in the Appendices.

DRIVER AND FLEET SAFETY

¶9-800 Managing driver and vehicle fleet safety

For couriers, delivery drivers, sales representatives, drivers of trucks, taxis and buses, as well as countless others for whom driving is an incidental part of the job, the vehicle's cabin and the open road are part of the working environment. This means in practice that while employers have less control over road safety than they have over the safety of their own premises, they can and should exert their influence to enhance driver safety through a range of measures including vehicle selection, maintenance, fleet management and driver training.

Control of the fleet and driver safety will involve several people, such as a purchasing manager, training coordinator, fleet manager and maintenance staff, meaning that a health and safety officer may have to operate through several channels.

Some of the ways in which management can improve driver and fleet safety are summarised in the following table.

Summary — driver and fleet safety

1. *Policy* — Management could emphasise road safety in its health and safety policy, or in a separate program, setting out areas of responsibility and accountability.
2. *Driver selection* — The recruitment process could assess factors such as experience on other jobs, attitude towards safety, past driving record, references, licence check, medical examination, knowledge of vehicles, maintenance and road rules, and a road test.
3. *Driver training* — The induction process could emphasise road safety (including issue of program and traffic rules). The importance of customer relations as well as personal and vicarious liability could also be mentioned. Instruction courses could be given in defensive driving, emergency procedures, reporting accidents/breakdowns, and how to check vehicles for safety. Employers should be aware that training in defensive driving can induce a false sense of security, and this should not be relied upon as a sole strategy.
4. *Vehicle design and selection* — Purchasing decisions should take into account features such as cargo barriers, airbags, comfortable seating and other safety features, as well as price, fuel consumption and other matters.
5. *Driving schedules/fatigue* — Fatigue is believed to contribute to a significant proportion of fatal crashes in heavy vehicles. Driving schedules designed to improve fatigue management can help control the risks. Drivers should not be given financial incentives which will motivate them to spend an excessive number of hours behind the wheel, and contract arrangements should not force drivers to accept dangerously fatiguing schedules.
6. *Mobile phones* — Use of mobile phones while driving can cause accidents because the driver is distracted while attending to the phone and/or the phone conversation. The organisation's policy on driving safely should specify that the driver should pull over and stop before making phone calls. Driving while talking on a hand-held mobile phone is illegal in all states and territories in Australia.

7. *Speed* — As speed is often a critical factor on road accidents, some organisations have chosen to use speed delimiters. Together with better scheduling and driver training, they can help reduce risks.
8. *Maintenance and record keeping* — The maintenance arrangement may depend on the agreement with the vehicles' lessors. Records must clearly and efficiently show when maintenance is needed, what maintenance is required, record of completion, costs, vehicle condition reports and history cards, accident records (causes and circumstances) and driver records (tests, training, appraisals, complaints, accidents). Spot checks of vehicles could also be made periodically. Any breaches of set standards could be penalised in some way, such as by loss of the private use of the vehicle for a set period.

Further references

StateFleet Driver Safety, NSW Department of Commerce. (Website at http://www.statefleet.dpws.nsw.gov.au/Driver+Safety/Driver+Safety.htm)

Transport Industry Guide to Occupational Health and Safety, WorkSafe Victoria, 2005.

Tyre Fitting Safety, WorkSafe Victoria, 2004.

Driver Safety. The website at http://web.info.com/infocom.us2/search/web/driver%20safety?CMP=2565&itkw=driver%20safety provides a large number of items on many different aspects of driver safety.

See the websites of government OHS authorities, road transport departments and other information providers such as those listed in the Appendices.

Chapter 10

MANAGING RISKS TO HEALTH AND WELLBEING

SMOKING AND THE WORKPLACE

¶10-000 Overview

Occupational diseases are illnesses or conditions caused by exposure to health hazards at work. In some cases the damage to health and wellbeing occurs gradually, as in noise-induced hearing loss, overuse injuries or other musculoskeletal disorders. In other cases, however, the damage is immediate, for example, if a health care worker is accidentally jabbed with a syringe containing infected blood, or workers are overcome by a toxic gas leak.

Exposure to chemicals or other hazardous substances such as pesticides has caused a great deal of sickness and death, but there is often a long delay between the workplace exposure which caused the problem, and the appearance of symptoms. In the case of work-related cancer, the long delay between cause and effect has often meant that neither the sick person nor the medical practitioner realised that the disease was actually work related, and therefore the case was never counted amongst workers compensation statistics. For this reason, available statistics tend to underestimate the real extent of work-related health problems.

The National Occupational Health and Safety Commission has estimated that over 2,000 Australians could be dying every year from occupational diseases, mainly due to workplace exposure to hazardous substances. Though exact figures are not available, it is likely that work-related disease is a more significant problem than work-related injury, and may even be responsible for more deaths than road accidents.

This chapter outlines the most notable risks to health and wellbeing at work, but it is by no means exhaustive. Some examples of occupational diseases not covered in this chapter are:

- diseases acquired from animals. In recent years, the risk of infectious diseases such as avian influenza (bird flu) has attracted widespread attention, but animal diseases that can affect human beings have a long history, and are well known as an occupational hazard for people whose work brings them into contact with animals
- hepatitis or other blood borne diseases that health workers can acquire from patients
- contagious diseases such as rubella or cytomegalovirus which childcare workers can acquire from the children they care for, and which can affect the health of the unborn child if the worker is pregnant
- lung diseases which make a gradual appearance, such as scarring of the lungs caused by long-term breathing of dust containing asbestos or silica (from sandstone). Work-related respiratory problems have taken a heavy toll on the health of Australian workers, and the effects are still being felt by thousands of workers around the country.

Because the risks of work-related health problems are often less obvious than the risks of accident or injury, the process of identifying, assessing and controlling the risks needs to take this into account. Hazard identification procedures therefore need to include investigation of available information from government departments and other sources such as industry and unions, to make sure systems of work provide protection from disease, as well as injury.

MANUAL HANDLING INJURIES

¶10-120 Scope of the problem

Injuries resulting from the manual handling of loads comprise the largest single category of workplace injuries, typically accounting for at least one-third of all injuries.

There is a strong association between manual handling and work-related back pain. About 70% of back injuries are strains and sprains resulting from manual handling, though back pain can often be caused by slips, trips and falls. Back pain is the most common type of manual handling injury, though injuries to the shoulders, neck, knees and ankles can also result from manual handling. These types of injuries are collectively known as "musculoskeletal disorders", and they can occur either suddenly or over a long period of time, by a gradual accumulation of damage. Occupational overuse disorders are related in terms of the types and causes of injury, and the types of strategies which can be used to address the problem. See ¶10-620–¶10-640 for more information about occupational overuse syndrome.

Manual handling is defined by the National Occupational Health and Safety Commission's *National Standard for Manual Handling* as "any activity requiring the use of force exerted by a person to lift, lower, push, pull, carry or otherwise move, hold or restrain any animate or inanimate object".

As well as the national standard, the NOHSC has also produced a *National Code of Practice for Manual Handling*. State and territory legislation is based on this standard and code of practice (see ¶10-150).

The term "manual handling injuries" refers to a number of conditions. Sprains, strains, and torn muscles or ligaments are the most common injuries. Injury to spinal discs is another more serious back condition. The effect of the injury and the type and location of the pain vary greatly depending on how far up the back the injury occurs.

Back injuries or other manual handling injuries may be particularly distressing to the worker involved. There is usually no visible sign of the injury, yet it causes a great deal of pain, severely limits the worker's home life, and may drag on for many months. Because there is no visible sign of injury, some tend to be labelled as "malingerers", which further aggravates the feeling of loss by the employee. Twenty per cent of workers with back injuries do not return to work within 60 days, and back injuries account for a third of all long-term compensation claims.

Manual handling injuries are sustained most frequently when lifting or setting down loads. An injury may be caused by a single lift, or as a cumulative effect of years of moving loads. Other situations that may lead to manual handling injuries include:

- carrying, pushing or pulling loads
- operating levers or other mechanical devices
- maintaining awkward posture over a long period, or while exerting force, and
- twisting or bending the spine, particularly while exerting force.

Personal factors such as poor levels of fitness, typical sleeping or sitting postures, cold muscles, tension or even the types of physical exercise engaged in can make people more vulnerable to a manual handling injury.

¶10-130 Identifying and assessing risk areas

The first step in reducing manual handling injuries at a workplace is to identify those jobs/tasks that could cause such injuries. Work methods and human characteristics also need to be taken into account. The following table lists factors for consideration.

Job/task factors	Work methods	People
• workplace layout	• trained lifting	• health
• frequency/duration	• team lifting	• physical capacity
• environment	• work flow	• skill
• force (weight)	• organisation	• experience
• nature of load	• accident reporting	• age
• type of handling	• personal protection	
• mechanical assistance		

While it has been traditional to concentrate on the weight of an object being handled, that is only one factor in determining whether a particular manual handling task presents a risk. A detailed checklist of risk factors is provided below. It is advisable to consider all the matters referred to when assessing manual handling risks. In summary, the major risk factors are:

- weight of the object (or force applied)
- work posture (distance of the object from the body and position of the body in relation to the object)
- frequency of lifting
- duration of lifting, and
- preparation for the task, ie warm-up.

The characteristics of the person performing the job/task will also influence the likelihood of injury. Factors such as gender, age, strength and history of back injury are known to affect the risk of injury.

In the following checklist, a positive answer indicates the possibility of a problem. For example, if there is a "Yes" answer to the question "Is the operator required to bend over frequently at low working heights?", this indicates a risk that may need to be addressed when control strategies are considered.

Manual handling risk factors — checklist

- Does the operator sit or stand constantly with little opportunity to move around?
- Is the operator required to bend over frequently at low working heights?
- Are extended reaches required by the operator to perform a task? If so, are these tasks sustained for at least one minute continuously or on a repetitive basis?
- Is the operator required to undertake work above shoulder level? If so, is this working posture sustained for at least one minute or on a repetitive basis?
- Is the operator required to undertake work in any of the following awkward postures or conditions?
  - (a) crouching while doing a task for longer than one minute
  - (b) twisting while undertaking manual handling
  - (c) handling of large-sized loads such as boxes or cases of more than 500 mm (18 inches) in any dimension
  - (d) handling of sheet materials or other large-sized loads without straps, special holders, or a second person
  - (e) handling of loads that can shift suddenly, such as live animals, liquids or bags of loose material
  - (f) handling of loads where the load is not shared equally between both hands or lifted single-handed, resulting in unbalanced posture
  - (g) sustained (more than 30 seconds) pushing and pulling of heavy loads
  - (h) pushing or pulling an object across the front of the body
  - (i) handling tasks that require an operator to bend to one side to lift an object or exert a force, resulting in an unbalanced posture
  - (j) inadequate clearances for moving the feet while doing a task that may result in an unbalanced posture or twisting
  - (k) a fixed workplace where it is not possible to adjust the location or heights of equipment being used, and/or
  - (l) simultaneous actions, especially with static muscle loading (ie unsupported limbs or application of force without any movement of limbs).
- For a seated operator, is there sufficient lumbar support (support for the lower back)?
- Is the operator required to undertake occasional lifting or lowering of objects greater than 16 kg (35 lb)?
- Is the operator required to undertake any frequent lifting or lowering of objects greater than 16 kg (35 lb)?
- Is the operator required to lift very heavy objects (ie over 55 kg (120 lb))?
- Does the operator lack suitable protective clothing for use during handling operations?
- Are suitable mechanical aids available for use during handling operations? (See section on "Mechanical aids" at ¶10-140.)
- Does the operator have appropriate training for the manual handling task?
- Do any of the following work environment factors make manual handling operations more difficult?
  - (a) thermal environment

(b) noise levels
(c) airborne contaminants including dusts, fumes, gases and vapours
(d) lighting glare (direct or indirect), and/or
(e) vibration levels.

- Does the operator handle objects with protrusions and/or sharp edges without mechanical assistance or appropriate protection?
- Do physical difficulties (eg excessive fatigue) arise from the time taken and/or distance travelled while undertaking the handling operations?
- Are there occasions where loads, while being handled, obstruct the operator's field of view?
- Do manual handling activities occur in work areas that have unduly rough floor surfaces, or floors that slope significantly or are slippery?
- Have strain or sprain injuries occurred while individuals were engaged in team lifting?

Sitting workplaces

- Do operators experience difficulty reaching or handling needed items within the seated work space?
- Do items being handled require hands to work at an average level of more than 15 cm (6 inches) above the work surface?
- While handling objects in the seated position, is large force exerted without mechanical aids?
- Are fine assembly or writing tasks done for a majority of the shift time?

Standing workplaces

- If objects weighing less than 4.5 kg (10 lb) are handled, could the operation be more suited to a varied work posture that allows change between sitting and standing?
- Are manual handling operations physically separated, requiring frequent movements between workstations?
- Are downward forces exerted, such as in packing operations? If so, can these forces be eliminated by the use of mechanical aids?
- Could suitable floor mats help to reduce the discomfort of operators whose job requires them to stand throughout the shift?

Sit/stand workplaces

- Could more variety in posture be introduced to jobs where multiple tasks are performed, some best done sitting and others best done standing?

¶10-140 Preventing manual handling injuries

Preventing back injuries and other musculoskeletal disorders means, as far as possible, eliminating the risks involved in manual handling tasks. Risk control can be addressed in three major ways:

- job redesign
- provision of mechanical aids, and
- training.

Of these three alternatives, job redesign is the preferable option. It can often be used to eliminate manual handling tasks, or to alter them so that they are not stressful on the back.

Job redesign

Whilst job redesign must be considered in the light of the individual job/task, it is possible to provide some examples.

In order to reduce bending, it may be possible to raise the working level. Adjustable height work-benches may solve the problems of different workers using the bench, or different tasks being carried out on the same bench. Materials or tools that will be used at the working level should not be stored at a lower level.

Twisting can be reduced by providing all tools and controls in front of the worker, or by allowing workers sufficient space to move their whole body.

The risks of lifting can be decreased by reducing the size of loads to be lifted, or by increasing the size of loads so that they must be lifted mechanically. The most commonly used materials and the heaviest materials should not be stored at ground level or above shoulder level.

Mechanical aids

Mechanical aids to manual handling tasks include:

- hooks
- crowbars
- trolleys
- belt conveyors or roller conveyors
- jacks
- cranes and hoists
- forklifts and other industrial vehicles, and
- adjustable tables/platforms.

Training

Proper training in manual handling tasks should be provided regardless of other preventive strategies adopted. The training should encourage workers to consider better ways of performing tasks (job redesign) as well as covering handling techniques. Workers should be made aware of the risks involved with each job/task, and the way of performing that job/task with least risk.

Where mechanical aids are used, it will be necessary to provide training in the proper use of the equipment.

There has been some debate over the best lifting technique. However, the following principles are generally recognised as important when training people in safe manual handling:

1. *Plan.* The person undertaking the manual handling should assess the load and determine how it will be handled and where it will be placed as

one way of avoiding over-exertion injuries. At this stage, the person can assess whether a handling aid or the assistance of another person is needed.

2. *Determine the best technique.* All factors should be taken into account when determining the best technique. The person should do the manual handling efficiently and rhythmically. The best technique involves suitable balance and avoidance of unnecessary bending, twisting and reaching. Above all else, the manual handling technique must be specific to the requirements of the task.
3. *Take a secure grip on the object being handled.* The grip helps determine how safe the task will be. Wherever possible, a comfortable power grip (using the whole hand) should be used, rather than a pinch grip (the thumb and fingers only).
4. *Pull the load close to the body.* In lifting in particular, it is important to have the centre of gravity of the load close to the body. This prevents excessive stress on the back, and makes the strongest muscles of the arms available to hold the load. It is important to minimise the effects of acceleration by handling the load slowly, smoothly and without jerking.
5. *Vary handling tasks with lighter work.* The job should be designed to provide alternative tasks that do not heavily stress the same muscles.

¶10-150 Regulation of manual handling

Employers need to assess all manual handling risks in their workplace, and then use job redesign, mechanical aids and training to reduce those risks. This is the approach taken by all states and territories that have manual handling legislation consistent with the National Commission's *National Standard and Code of Practice on Manual Handling*.

Legislation and codes of practice are in place in all jurisdictions:

NSW

Occupational Health and Safety Regulation 2001

Vic

Occupational Health and Safety (Manual Handling) Regulations 1999

Code of Practice for Manual Handling

Code of Practice for Manual Handling in the Furniture Removal Industry

Qld

(Code of Practice) Manual Tasks Advisory Standard 2000

(Code of Practice) Manual Tasks Involving the Handling of People Advisory Standard 2001

SA

Occupational Health, Safety and Welfare Regulations 1995

Code of Practice for Manual Handling

WA

Occupational Health and Safety Regulations 1996

Code of Practice for Manual Handling

Code of Practice — Manual Handling in the Building Industry

Code of Practice for Manual Handling — the Handling of People

Tas

Workplace Health and Safety Regulations 1998

NT

Work Health (Occupational Health and Safety Regulations) 1992

Approved Code of Practice for Manual Handling

ACT

Occupational Health and Safety (Manual Handling) Regulations 1998

Code of Practice and Standard on Manual Handling

Further references

(Note that some of these references are no longer available from the organisations that produced them but may be found in libraries with special OHS collections. See also the websites of government OHS authorities and other OHS information providers such as those listed in Appendices A and B.)

National Standard for Manual Handling [NOHSC:1001(1990)], National Occupational Health and Safety Commission.

National Code of Practice for Manual Handling [NOHSC:2005(1990)], National Occupational Health and Safety Commission.

The Hands On Guide: Manual Handling, CCH Australia Limited, 2000.

Core Training Elements for the National Standard for Manual Handling, National Occupational Health and Safety Commission, 1995.

Kroemer, K and Grandjean, E, *Fitting the Task to the Human. A Textbook of Occupational Ergonomics*, 5th edn, Taylor and Francis, 1997.

Mital, A, Kilbom, A and Kumar, S, *Ergonomics Guidelines and Problem Solving*, Amsterdam, Elsevier, 1999.

Guidance Note for Manual Handling in the Retail Industry, National Occupational Health and Safety Commission, 1992.

McPhee, B, *Ergonomics for the Control of Sprains and Strains in Mining*, National Occupational Health and Safety Commission/Joint Coal Board, 1994.

Back Care at Work, WorkSafe Western Australia.

Ergonomic Workstations for Keyboard Operators, WorkSafe Western Australia.

It's Your Back — Handle with Care, WorkSafe Western Australia.

Backcare at Work Training Package, WorkSafe Western Australia.

Guide to Manual Handling in the Wholesale and Retail Industry, WorkSafe Western Australia.

Manual Handling, Work Health Authority (NT).

Guidance Note: Back Belts are not Effective in Reducing Back Injuries, WorkSafe Victoria, June 2005.

HAZARDOUS SUBSTANCES

¶10-220 Chemical hazards/hazardous substances

Thousands of different chemicals are used in industry, with new ones being introduced at a global rate of 1,000 to 2,000 per year. The properties and health effects of many of these chemicals are well known, but it is not possible to obtain comprehensive information regarding all industrial chemicals.

Examples of chemicals in common usage include solvents, alcohol, salt, detergents, paint, petrol, pesticides and medicines. Most chemicals are safe in the amounts normally used. If some are used excessively or improperly, however, they may result in disease, injuries or even death. Some chemicals may harm the health at low doses or at levels at which people may be reasonably expected to come into contact with them. These are known as "hazardous substances".

Hazardous substances are not limited to those substances commonly thought of as "chemicals". For example, asbestos, sandstone dust (containing silica), and vehicle emissions can all be hazardous substances. Similarly, fumes from various sources are hazardous substances, as are dusts encountered in chicken sheds and some other agricultural facilities. The asbestos disaster reminds us that the health effects of some substances may not become apparent until after many years of exposure.

However, it is possible to minimise the risks associated with chemicals and other hazardous substances at the workplace by adopting certain procedures. These procedures revolve around two aspects:

1. adequate information, readily available, and
2. control measures to eliminate or minimise the risk to workers.

These two aspects are developed in the following paragraphs.

The main hazards from chemicals are:

- environmental pollution
- fire/explosion, or other physical damage, and
- health effects.

This commentary will concentrate on the health effects of hazardous substances at the workplace, with some discussion of precautions against emergencies such as fires (see ¶10-260). However, an organisation planning its hazardous substances management strategy will have to take all three forms of hazard into account.

The definition of a hazardous substance is a substance that meets either or both of the following criteria:

- It is listed in the National Occupational Health and Safety Commission's *Designated List of Hazardous Substances.*
- It meets the requirements of the National Commission's *Approved Criteria for Classifying Hazardous Substances.*

In general, the term "hazardous substances" is used to indicate the potential to harm health — with the harm being done over time. In this way, "hazardous substances" are distinguished from "dangerous goods". The latter are generally assumed to pose an immediate risk to personnel, property or environmental safety, in terms of explosion, fire or poisoning. There is significant overlap between these two categories, such that many hazardous substances are also classified as dangerous goods. Separate legal requirements apply to dangerous goods; and where an overlap occurs, both the hazardous substances requirements and the dangerous goods provisions apply.

The situation is complicated by the existence of other chemical classifications as well, such as therapeutic goods and poisons. Chemical management is more difficult when different chemicals are regulated by different regulators (eg industrial, therapeutic, agricultural and veterinary, poisons), and different requirements apply with respect to labelling, provision of information, risk assessments, etc.

The Globally Harmonized System of Classification and Labelling of Chemicals (GHS), developed by the United Nations and adopted by member states in 2002, is a step towards resolving the complications caused by the overlap of the various systems of regulation. The GHS is a system that harmonises the classification of hazardous substances and hazard communication. It does not distinguish between hazardous substances and dangerous goods, contrary to what is now the case in all Australian jurisdictions (except in the ACT with its recently revised *Dangerous Substances Act 2004*). By harmonising chemical management on a worldwide basis, the GHS will hopefully facilitate trade, enhance protection of human health and the environment, and reduce duplication in the assessment process. The GHS has a target implementation date of 2008.

National chemical assessment scheme

The National Industrial Chemicals Notification and Assessment Scheme has been in operation since 1990. Under the scheme, all new and some existing chemicals imported into or manufactured in Australia are required to be notified and assessed. Each chemical is assessed for its potential effects on occupational health, public health and the environment. The scheme is administered by the federal Department of Health and Ageing.

¶10-230 Health effects

Different chemicals cause a wide range of health effects.

Acute poisons are those where a short exposure causes an immediate health effect. Some chemicals, such as strong acids, have an immediate effect on contact with the skin or eyes, while others, such as cyanide, have a rapid effect as soon as they are inhaled.

Chronic toxins are chemicals that cause harm due to prolonged exposure. For example, build-up of mercury in the body can cause nerve and brain poisoning. Note that many chronic toxins will become acute poisons in large doses.

Corrosive chemicals and irritants usually affect mucous membranes (mouth, nose, throat, lungs). Inhalation of a high dose of sulphuric acid fumes will cause lungs to collapse and possibly death.

Allergens or sensitisers cause a change in the body's reaction to that substance, so that subsequent exposure causes a reaction, most notably dermatitis or asthma. For example, isocyanate compounds can cause sensitisation resulting in asthma-like symptoms. It is not usually possible to tell who will become sensitised to a particular chemical. Occupational skin problems, often caused by allergens or sensitisers, are discussed at ¶10-420 et seq.

Carcinogens are those chemicals that are capable of causing cancer. They include asbestos, synthetic mineral fibres, arsenic compounds, benzene and vinyl chloride. Occupational cancer is discussed in more detail at ¶10-320 et seq.

Mutagens change the genetic code of cells and can cause mutations.

Teratogens affect the growth of the fertilised egg and embryo, also causing mutations.

Routes of entry

In industry, there are three main ways in which chemicals can enter the body.

1. *Inhalation*: breathing chemicals into the lungs; the most common method of chemicals entering the body at work.
2. *Body contact*: chemicals coming into contact with the skin or eyes should generally be avoided.
3. *Ingestion*: eating or drinking; usually not such a problem, but hands should be washed before eating or smoking to avoid accidental ingestion.

Monitoring doses

The risk of adverse health effects depends on the dose of a chemical that is absorbed into the body. For this reason it is important to monitor the concentration of dangerous substances in workroom air. These measurements require special equipment as well as a knowledge of occupational hygiene (this has nothing to do with personal hygiene — see "The occupational hygienist" in Appendix C). The tests must be done in areas where workers are located, and they must be done so that both typical and worst levels are recorded.

These measurements can then be compared with exposure standards determined by various health and safety bodies. Exposure standards establish a level of exposure to which, it is believed, most workers may repeatedly be exposed without adverse effect. The NOHSC issued such standards for more than 650 chemicals.

One commonly used exposure standard is known as a *Threshold Limit Value* (TLV). Often a TLV is issued as a *Time Weighted Average* (TLV-TWA), which refers to the safe limit of exposure averaged over eight hours. This can be exceeded for a short period, provided the exposure does not exceed the Short-Term Exposure Limit (TLV-STEL), which usually refers to the safe limit for 15 minutes of continuous exposure. Some dangerous chemicals will not have a Time Weighted Average, but a Ceiling TLV, which should not be exceeded at any time.

¶10-240 Management of hazardous substances

The first step in any hazardous substances management program is to establish an inventory of all chemicals used at the workplace. Where this is not already in place, it will involve a survey of the chemicals and other hazardous substances in the workplace, as well as establishing a procedure for recording new substances as they enter or are produced by the work process at the workplace. The questions to be asked include:

- What materials are present?
- Where are they located?
- How are they stored?
- How much is stored at each location?
- How are they used?
- How often are they used?

All areas of the workplace must be considered, including laboratories, receiving areas, offices and maintenance sections. Hazardous substances not purchased as "chemicals" must also be included in the inventory. For example, dust at a minesite is not stored or used, but is present in the work environment and has potential health effects which must be managed.

Material safety data sheets

Once it is known which chemicals are present at the workplace, it is necessary to obtain information on those chemicals. This can be done by obtaining a material safety data sheet (MSDS) for each chemical. These data sheets detail relevant health and safety information on a substance. It should be possible to obtain an MSDS from the supplier or manufacturer of the substances. If the business is producing or generating a hazardous substance, it may be necessary to compile an MSDS for it.

The NOHSC has published a *National Code of Practice for the Preparation of Material Safety Data Sheets*, which should be used when compiling MSDSs. Legislation requires that suppliers and manufacturers of substances must supply this information. It is worthwhile insisting that all materials coming into a workplace are supplied with an MSDS, or at least with a list of ingredients and safety information. However, it will then be necessary to "customise" the MSDS, by adding information such as the use to which the substance is put, and the place(s) where it is stored. Safety hints discovered when using the substance should also be added to the MSDS, so that they are not lost when new personnel come into the workplace.

Additional information for completion of MSDSs can be obtained from health and safety bodies.

The list of chemicals stored at the workplace and the attached MSDSs should be stored in a filing system with easy access. Inexpensive software is available to perform this task.

The following sample MSDS is reproduced with permission of the Chamber of Minerals and Energy of Western Australia, Inc.

Labelling

Another important aspect of hazardous substances management is adequate labelling. It is not possible to provide all safety information on a label, and it is better if the label contains minimal information but enables the user to refer easily to the correct MSDS. A clear substance name is the major function of a label, with perhaps an identifying number and mention of any particularly dangerous ingredients. Labelling should conform with the NOHSC's *Code of Practice for the Labelling of Workplace Substances*.

MATERIAL SAFETY DATA SHEET

Product	Noricene 8940

Manufacturer Details: Nalco Australia Pty Ltd
[Address, telephone, Fax]

SYNONYMS: None

UN: 1789	**CLASS 8**	**SUB RISK:**	**HAZ CHEM CODE:** 2R
EPG: zz38	**PACKAGING GROUP:** II		POISON SCHEDULE: 6

USES: Inhibited hydrochloric acid descaler.

PHYSICAL AND CHEMICAL PROPERTIES

APPEARANCE: Clear amber liquid with a pungent odour.

BOILING POINT: 84°C	**MELTING POINT:** 0°C
VAPOUR PRESSURE: 35 mmHg	**VAPOUR DENSITY:** 1.15
SPECIFIC GRAVITY: 1.27	**SOLUBILITY (WATER):** Soluble
FLASH POINT: None	**EXPLOSION LIMITS:** None
% VOLATILES: Negligible	**pH:** 1% Solution: 1

INGREDIENTS

Chemical Name	CAS Number	Proportion
Hydrochloric acid	7647-01-0	44%
Proprietary corrosion inhibitor		<10%

HEALTH HAZARD INFORMATION

INGESTION: Corrosive, may cause burning of the mouth, throat and stomach. Ingestion may cause coughing, constriction of the throat due to swelling of the larynx and vomiting of blood or blood in diarrhoea. Breathing difficulties, shock and convulsions may follow.

EYE: Vapour is severely irritating. Liquid is severely irritating and may cause conjunctivitis, ulceration and corneal burns. Permanent eye damage may result.

SKIN: Irritating after brief contact. Prolonged contact may result in inflammation and burns.

INHALATION: Product has relatively low volatility so inhalation of hazardous quantities of vapour is unlikely to occur during normal use. However, if generated, inhalation of vapours or spray mists are severely irritating. Inhalation may cause burning sensations, higher concentrations may cause coughing, choking and difficulty in breathing due to oedema of the lungs. Symptoms of oedema may have delayed onset.

CHRONIC: Hydrochloric acid
LCLo (inhaled, human): 1300 ppm/30M
LD50 (oral, rabbit): 900 mg/kg
LCLo (inhaled, rabbit): 4413 ppm/30M

Inhalation and ingestion are the routes of entry into the body. Adverse health effects are related to corrosive nature. The estimated fatal dose for an adult is 1 mL of concentrated acid.

FIRST AID PROCEDURES

INGESTION: NEVER GIVE AN UNCONSCIOUS PERSON ANYTHING TO DRINK NOR ATTEMPT TO INDUCE VOMITING. If the person is conscious, rinse mouth out with water ensuring that mouth wash is not swallowed. Give about 250 mL (2 glasses) of water to drink. DO NOT attempt to induce vomiting. Seek URGENT medical attention.

EYE: IMMEDIATELY hold eyelids open and rinse the eye continuously with a gentle stream of clean running water for at least fifteen minutes. Seek URGENT medical attention. If practical, continue irrigation of the eye during transportation to medical facilities.

SKIN: Remove contaminated clothing. Rinse the affected area with water then wash thoroughly with soap and water. Use water alone, if soap is unavailable. Seek medical attention if any soreness or inflammation of the skin persists. Launder affected clothing before re-use.

continued over ...

INHALATION: Remove to fresh air. Keep warm and at rest. If breathing is laboured, hold in a half upright position (this assists respiration). Apply artificial respiration if breathing has stopped. Seek URGENT medical attention for all but the most minor cases of over-exposure.

PRECAUTIONS FOR USE

ENGINEERING CONTROL: Ventilation requirements depend on the concentration of solution in use, the quantity and the method of application. General (mechanical) ventilation is adequate for minor use. Local exhaust ventilation may be required if handling large quantities of concentrated product.

FLAMMABILITY: Non-combustible. May evolve acrid fumes if heated. Concentrated solutions react vigorously with alkalis. May evolve hydrogen on contact with metals.

PERSONAL PROTECTION: Requirements are dependent on working conditions, quantity of product in use and its concentration. For minor use, acid resistant safety goggles and rubber gloves may be adequate. If concentrated solutions or large quantities are used, [wear] acid resistant safety goggles, face shield, gloves or gauntlets, overalls and splash apron. A half face respirator with acid gas filter is required if the product is being sprayed or handled in such a way that hydrogen chloride vapour is generated. In confined spaces use air supplied breathing apparatus. N.B. TAKE THE LIMITS OF ABSORPTION CAPACITY INTO ACCOUNT. CHANGE FILTERS REGULARLY.

EXPOSURE STANDARD

E.S.	**TWA**		**STEL**	
CHEMICAL NAME	**ppm**	**mg/m³**	**ppm**	**mg/m³**
Hydrochloric acid	5.00	7.50		Peak

T.L.V.	**TWA**		**STEL**	
CHEMICAL NAME	**ppm**	**mg/m³**	**ppm**	**mg/m³**
Hydrochloric acid			5.00	7.50

SAFE HANDLING PROCEDURES

STORAGE TRANSPORT: Store in a cool well ventilated area. Avoid contact with reactive metals. Keep separate from alkaline substances. Protect containers against physical damage.
Class 8 Corrosive Substances should not be transported or stored with goods of:

Class 1	Explosives
Class 4.3	Dangerous When Wet Substances
Class 5.1	Oxidising Agents
Class 5.2	Organic Peroxides
Class 6	Poisonous (toxic) Substances (where the poisonous substances are cyanides and the corrosives are acids)
Class 7	Radioactive Substances

Foodstuff and foodstuff empties

SPILLS DISPOSAL:
SPILLS: Dilute solutions and very minor spills may be flushed away with water. For major spills or those involving concentrated product wear full protective equipment including air supplied breathing apparatus, gloves, acid resistant overalls and rubber boots (trousers should be worn OVER boots, not tucked in). If possible the spill should be contained by damming with earth or sand and then covered with a weak alkali such as soda ash or sodium bicarbonate. It can then be flushed away with copious quantities of water.
DISPOSAL: Dilute solutions: flush to drains with plenty of water. Concentrated solutions: dilution and chemical reaction or to approved land-fill.

FIRE EXPLOSION: Non-combustible but may evolve acrid fumes in a fire situation. May evolve hydrogen, a highly flammable gas, on contact with reactive metals.
Wear self contained breathing apparatus. Extinguish using whatever is suitable for the primary cause of the fire. Water sprays may be used to 'knock down' vapour.

USER COMMENTS

¶10-250 Control of hazardous substances

There are various methods of controlling chemical hazards at the workplace. These can be split into two basic groups:

1. engineering and design methods, and
2. controls which depend on modification of worker behaviour.

Engineering and design methods are more effective in reducing chemical hazards and should be considered before behavioural controls.

Engineering and design controls

1. *Elimination or substitution.* Where possible, hazardous substances should be eliminated from the particular process or substituted with a less dangerous chemical. It may also be possible to substitute a less dangerous process involving the same chemical, such as using dipping rather than spraying. A good technical knowledge of the particular industry/process is required to determine whether elimination or substitution is possible.
2. *Isolation or enclosure.* It may be possible to isolate the majority of workers from a dangerous process or, alternatively, to enclose the process completely. For example, many solvent degreasing units are designed so that the degreasing solvent is completely enclosed.
3. *Ventilation.* This can be used to control airborne contaminants. Local exhaust ventilation, where the contaminant is sucked away from the point of release, is the most effective. Dilution ventilation relies on clean air being pumped into the workplace and the old air being released. Natural ventilation is the least effective method.

Controls that rely on human behaviour

1. *Information and training.* Workers should understand the hazards that they face, be able to access and interpret MSDSs and know the procedures they should follow to minimise the risks.
2. *Supervision.* Good supervision is necessary to ensure that work is done safely, and problems are addressed as they arise. Supervisors need to be trained to recognise unsafe systems of work.
3. *Monitoring.* Monitoring is necessary to ensure that particular risks are under control. Monitoring is discussed above in the context of exposure standards (see ¶10-230). Biological monitoring, that is, testing blood or urine samples to determine levels of exposure to hazardous substances, is another method appropriate under some circumstances for checking whether risks are adequately controlled.
4. *Personal protective equipment.* This is the least preferred option. Personal protective equipment (eg gloves, respirator, protective clothing) should be used only where it is not possible to introduce engineering or design controls. Protective equipment is discussed in more detail at ¶3-140.

Legislative provisions, codes and standards

In 1994 the NOHSC declared a national regulatory reform package for the control of workplace hazardous substances. The package included *National Model Regulations for the Control of Workplace Hazardous Substances*, as well as codes of practice for the preparation of MSDSs, for the labelling of workplace substances, and for the control of workplace hazardous substances. The NOHSC also published a code for the control of scheduled carcinogenic substances.

Since then, all states and territories have adopted hazardous substances regulations consistent with the National Commission's model regulations and codes of practice.

¶10-260 Emergency planning

A final consideration in relation to chemical safety is planning for emergencies. This relates particularly to substances regarded as dangerous goods. It involves planning for the safest storage of chemicals, and placarding dangerous goods in order to be of assistance to emergency services.

When planning a new installation, it is worthwhile to consider the sorts of emergencies that may arise. Flammable chemicals, particularly, should be stored separately from the main workplace.

Placarding

All chemical storage areas should be placarded in order to provide information to emergency services. This means placing clearly visible placards setting out the name of the substance, any standard safety signs (flammable liquid, flammable solid, poison gas, oxidising agent, corrosive, etc) and the Hazchem Code for that substance. Hazchem Codes are a system, developed originally by the UK Fire Brigade, for telling the fire services certain information such as the type of fire extinguisher that should be used, what sort of personal protection is necessary and whether evacuation should be considered.

The NOHSC published a *Guidance Note for Placarding Stores for Dangerous Goods and Specified Hazardous Substances*, which details the sorts of placards that are necessary. Specific placards should be placed on each chemical in bulk storage, while composite placards (setting out the type of chemicals contained within) can be placed at the entrance to a mixed storage. Examples of placards are shown below.

Placarding is also required on dangerous goods that are being transported. These requirements are set out in the *Australian Code for the Transport of Dangerous Goods by Road and Rail* (ADG Code). The ADG Code has been adopted by all states and territories. The code applies to a large number of dangerous substances, which it splits into various classes depending on their properties.

Chemical plan/manifest

A plan of all hazardous chemicals stored at the workplace (sometimes referred to as a manifest) also assists emergency services. The plan should be kept in a readily accessible location, and should set out all the chemicals on site by class, quantity and location.

State/territory legislation and codes of practice

NSW

Occupational Health and Safety Regulation 2001

Code of Practice for the Control of Workplace Hazardous Substances

Code of Practice for the Preparation of Material Safety Data Sheets

Code of Practice for the Labelling of Workplace Substances

Vic

Occupational Health and Safety (Hazardous Substances) Regulations 1999

Qld

Workplace Health and Safety Regulation 1997

(Code of Practice) Hazardous Substances 2003

The Storage and Use of Rural Chemicals at Rural Workplaces (Industry Code of Practice 2000)

SA

Occupational Health, Safety and Welfare Regulations 1995

Code of Practice for the Control of Workplace Hazardous Substances

Code of Practice for the Preparation of Material Safety Data Sheets

Code of Practice for the Labelling of Workplace Substances

WA

Occupational Health and Safety Regulations 1996

National Code of Practice for the Control of Workplace Hazardous Substances

National Code of Practice for the Preparation of Material Safety Data Sheets

National Code of Practice for the Labelling of Workplace Substances

Tas

Workplace Health and Safety Regulations 1998

ACT

Approved National Model Regulation as a Code of Practice

National Code of Practice for the Control of Workplace Hazardous Substances

National Code of Practice for the Preparation of Material Safety Data Sheets

National Code of Practice for the Labelling of Workplace Substances

NT

Work Health (Occupational Health and Safety) Regulations 1992

National Code of Practice for the Control of Workplace Substances

National Code of Practice for the Labelling of Workplace Substances

National Code of Practice for the Preparation of Material Safety Data Sheets

Further references

Guidance Note for Placarding Stores for Dangerous Goods and Specified Hazardous Substances, [NOHSC:3009(1990)], National Occupational Health and Safety Commission.

Guidance Note on the Interpretation of Exposure Standards for Atmospheric Contaminants in the Occupational Environment 3rd Edition, [NOHSC:3008(1995)], National Occupational Health and Safety Commission.

National Model Regulations for the Control of Workplace Hazardous Substances, [NOHSC:1005(1994)], National Occupational Health and Safety Commission.

Guidance Note for the Assessment of Health Risks Arising from Hazardous Substances in the Workplace, [NOHSC:3017(1994)], National Occupational Health and Safety Commission.

National Code of Practice for the Labelling of Workplace Substances, [NOHSC:2012(1994)], National Occupational Health and Safety Commission.

National Code of Practice for the Preparation of Material Safety Data Sheets 2nd Edition, [NOHSC:2011(2003)], National Occupational Health and Safety Commission.

National Code of Practice for the Control of Workplace Hazardous Substances, [NOHSC:2007(1994)], National Occupational Health and Safety Commission.

National Code of Practice for the Control of Scheduled Carcinogenic Substances, [NOHSC:2014(1995)], National Occupational Health and Safety Commission.

National Code of Practice for the Storage and Handling of Workplace Dangerous Goods, [NOHSC:2017(2001)], National Occupational Health and Safety Commission.

Code of Practice for the Safe Removal of Asbestos, [NOHSC:2002(2005)], National Occupational Health and Safety Commission.

Guidance Note for the Control of Workplace Hazardous Substances in the Retail Sector, [NOHSC:3018(1994)], National Occupational Health and Safety Commission.

Approved Criteria for Classifying Hazardous Substances, [NOHSC:1008(2004)], National Occupational Health and Safety Commission.

List of Designated Hazardous Substances, [NOHSC:10005(1999)], National Occupational Health and Safety Commission.

In addition to the above, the NOHSC publishes a number of codes of practice and guides dealing specifically with the more common hazardous chemicals, including inorganic mercury, benzene, glutaraldehyde, ethylene dioxide, organic and inorganic lead, asbestos, vinyl chloride, timber preservatives, arsenic, cyanide, hydrogen fluoride, industrial organic solvents, foundry chemicals, isocyanates, solvent vapour degreasers, synthetic mineral fibres and welding fumes.

Controlling Hazardous Substances, ACTU Occupational Health and Safety Unit, 1995.

Chemicals Management at the Workplace, a Step by Step Guide, WorkSafe Victoria, 2002.

Hazard Alert: Managing Workplace Hazardous Substances, CCH Australia Limited, 1995.

Laboratory Safety Manual, CCH Australia Limited, 1990.

Standards Australia has produced many standards relevant to a particular hazard or situation.

AS 2030 *Gas Cylinders* series of standards, Standards Australia.

AS 2507 *The storage and handling of agricultural and veterinary chemicals*, Standards Australia.

HB 76 *Dangerous goods — Initial emergency response guide*, Standards Australia.

Standard for the Uniform Scheduling of Drugs and Poisons, Commonwealth Department of Health and Ageing, Canberra, AGPS, CDHA, published annually.

Gad, SC, *Regulatory Toxicology*, Routledge, New York, 2001.

See also the websites of government OHS authorities and other OHS information providers such as those listed in the Appendices.

OCCUPATIONAL CANCER

¶10-320 Scope of the problem

Occupational carcinogens, or cancer-inducing agents, can include chemical substances, physical agents and work processes.

Cancer may be described as a growth or tumour that tends to spread and to reproduce itself. Occupationally induced tumours are no different in type and nature from those arising from non-occupational factors.

Estimates of the proportion of cancer cases that are work related vary, and are complicated by the fact that lifestyle factors (away from the workplace) such as social class, personal habits, nutritional standards and area of residence, may contribute to the occurrence of cancer. An example is skin cancer, which is becoming recognised as job related in various outdoor work situations, although obviously it can also be brought on by leisure time spent in the sun. However, it is clear that certain substances and work processes are in themselves carcinogenic.

¶10-330 Identification

It is important to identify occupational groups at high risk. A good example is provided by asbestos workers. It is now well established that asbestos dust is a source of lung cancer and mesothelioma (cancer of the lining of the lung). However, the late response to the recognition of asbestos dust as a source of cancer has resulted in deaths and disease in all industrial countries.

Identification of cancer-inducing agents is not easy and is basically the concern of research scientists. As work processes change over the years, and as there is almost invariably a long period between initial exposure to a carcinogen and the final onset of disease, a study may not precisely identify the effect of a specific carcinogen used in a particular work process.

The NOHSC's national model regulations list prohibited carcinogenic substances (which are prohibited except for specified uses) and notifiable carcinogenic substances (which require notification of use to the relevant public authority). (See the following table.)

PROHIBITED CARCINOGENIC SUBSTANCES

Substance Name (Chemical Abstract Number)

2-Acetylaminofluorene (53-96-3)

Aflatoxins

4-Aminodiphenyl (92-67-1)

Amosite (12172-73-5) (brown asbestos) — except for removal and disposal purposes and situations where amosite occurs naturally and is not used for any new application.

Benzidine (92-87-5) and its salts (including benzidine dihydrochloride (531-85-1))

bis (Chloromethyl) ether (542-88-1)

Chloromethyl methyl ether (107-30-2) (technical grade which contains bis (chloromethyl) ether)

Crocidolite (12001-28-4) (blue asbestos) except for removal and disposal purposes and situations where crocidolite occurs naturally and is not used for any new application.

4-Dimethylaminoazobenzene (60-11-7)

2-Naphthylamine (91-59-8) and its salts

4-Nitrodiphenyl (92-93-3)

NOTIFIABLE CARCINOGENIC SUBSTANCES

Substance Name (Chemical Abstract Number)

Acrylonitrile (107-13-1)

Benzene (71-43-2) — when used as a feedstock containing more than 50% of benzene by volume.

Chrysotile (12001-29-5) (white asbestos) — when used for the manufacture of asbestos products.

Cyclophosphamide (50-18-0) (cytotoxic drug) — when used in preparation for therapeutic use in hospitals and oncological treatment facilities, and in manufacturing operations.

3,3'-Dichlorobenzidine (91-94-1) and its salts (including 3,3'-Dichlorobenzidine dihydrochloride (612-83-9))

Diethyl sulfate (64-67-5)

Dimethyl sulfate (77-78-1)

Ethylene dibromide (106-93-4) — when used as a fumigant

4,4'-Methylene bis (2-chloroaniline) (101-14-4) — MOCA

2-Propiolactone (57-57-8)

o-Toluidine (95-53-4) and o-Toluidine hydrochloride (636-21-5)

Vinyl chloride monomer (75-01-4)

¶10-340 Prevention and control

It is vital to determine which substances used at a workplace may be carcinogenic. Material safety data sheets should be consulted to discover the make-up of each substance and any known health effects of that substance. It may be necessary to check a list of chemicals used at the workplace against a list of known carcinogens, such as the NOHSC list provided at ¶10-330. All areas of the workplace should be considered.

Each carcinogen should be documented regarding measures to be taken if exposure occurs. Workers should be aware of and involved in this process. Affected areas should be designated clearly in some way.

The following checklist of preventive measures should also be observed.

1. Consider substitution with a non-carcinogenic substance wherever possible.
2. Minimise contact with the carcinogen, through work layout, equipment design, repair/maintenance, waste disposal, emergency procedures, washing facilities and decontamination units.
3. Avoid burdening workers with time limits and deadlines during hazardous operations.
4. Operating instructions should be written, expressed in simple and clear terms, and allow for no deviation or improvisation.
5. Where special clothing is necessary, clean replacement clothing should be available, with special provisions for collecting and laundering or disposal of contaminated clothing. Information on hazardous materials should be conveyed to the cleaning sources to avoid risk to their employees. Change-rooms should be arranged so that there is no contact between work clothes and ordinary clothes.
6. Partaking of food and drink, as well as smoking, should be prohibited within the area. Suitable precautions to avoid contamination in lunchrooms, etc, should also be taken.
7. Regular monitoring of equipment (for leakage) and spot sampling of the working environment should be undertaken and measurements recorded.
8. Personal sampling devices should be used regularly to measure specific worker exposure, especially when changes in work processes occur. Monitoring records should be kept. Each employee should be informed of personal results.
9. Workers should be kept constantly informed of the nature of the hazard and all safety precautions initiated to minimise exposure.
10. Attention to the selection of employees to work in hazardous situations is important, in terms of technical competence, ability to work in a team, personal hygiene, and ability to give and understand instructions. Use of outside contractors should be avoided, unless they specialise in that particular field, or are able to comply with organisational directives (which should be written into the contract).
11. Company medical/health staff must be informed of the hazard and of any accidents or similar incidents.

Further references

National Model Regulations for the Control of Scheduled Carcinogenic Substances, [NOHSC:1011(1995)], National Occupational Health and Safety Commission, 1995.

National Code of Practice for the Control of Scheduled Carcinogenic Substances, [NOHSC:2014(1995)], National Occupational Health and Safety Commission, 1995.

Guidelines for Laboratory Personnel Working with Carcinogens or Highly Toxic Chemicals, National Health and Medical Research Council, 1990.

Occupational Cancer, National Occupational Health and Safety Commission, 1992. (Website at www.nohsc.gov.au/PDF/Standards/OccupationalCancer.pdf)

¶10-350 Synthetic mineral fibres

Synthetic mineral fibres (or man-made mineral fibres) refers to a variety of fibres made from such things as rock, slag, glass and clay. Examples include fibreglass, glasswool, glass filament, mineral wool (or rockwool) and ceramic fibres. These fibres are used for thermal and acoustic insulation and are coming into increasing use, in part as a replacement for asbestos.

Concerns have been expressed that synthetic mineral fibres (SMFs) can be inhaled and lead to lung cancer in the same way as asbestos. While this has not been fully established, SMFs are believed to be less hazardous than asbestos fibres. Synthetic mineral fibre dust contains fewer respirable fibres than asbestos dust, and SMFs do not break down very readily into smaller and even more easily respirable fibres. Ceramic fibres may be the most hazardous, due to their having smaller diameters than other SMFs.

Employers and workers should minimise exposure to SMFs. The NOHSC has set a national standard of 0.5 respirable fibres per millilitre of air, as the maximum acceptable exposure to SMFs. Note that SMFs can be purchased with binders and de-dusting oils added, which may reduce exposure to SMF dust.

Exposure to SMF dust can be reduced by various engineering and administrative methods such as:

- obtaining SMFs in pre-cut and wrapped form
- providing localised dust extraction where it is necessary to cut SMFs
- cleaning up waste and packaging regularly, and
- segregating areas where SMFs are in use from other work areas.

Where dust exposure is likely to be significant, employees should wear a respirator approved for SMF dust; long-sleeved, loose-fitting clothing and gloves (to prevent skin irritation); and eye-protection and head covering (for overhead work).

Codes of practice

The *National Code of Practice for the Safe Use of Synthetic Mineral Fibres* has been adopted by South Australia, Western Australia, the Northern Territory, the Australian Capital Territory and New South Wales.

¶10-360 Skin cancer

Occupational skin cancer is emerging as a serious problem in Australia. Australia has the highest rate of skin cancer in the world, with two out of three Australians who live to the age of 75 developing some type of skin cancer. Obviously it is not possible to attribute all sun damage to outdoor work, with a mixture of work and non-work exposure being responsible in each case. However, outdoor workers should ensure that they are adequately protected against sun damage to minimise their chances of developing skin cancer.

Skin damage from the sun is caused by ultraviolet (UV) radiation. In addition to outdoor workers, there are a number of occupations where workers may be subject to artificial sources of UV radiation. Note also that UV radiation passes through water and is reflected very efficiently from snow.

The following list sets out those occupations with potential exposure to ultraviolet light (UVL). Where the source of the UV radiation is not sunlight, it is noted.

- agriculturists
- animal farmers
- armed personnel
- barbers (artificial germicidal UVL)
- bricklayers
- construction workers
- cosmeticians (artificial germicidal UVL)
- council workers
- fishermen and women
- gardeners
- gas, water, sewage and drainage pipeline workers
- geologists
- greenkeepers
- laboratory technicians (artificial germicidal UVL)
- lifesavers
- lithographers (artificial UVL used in drying and curing)
- loggers and timber workers, tree fellers and similar
- maintenance workers (artificial UVL arc welding equipment)
- oilfield workers
- open-cut miners
- paint processors (artificial UVL used in drying and curing)
- pipe cutters (artificial UVL arc welding equipment)
- pipeline workers (artificial UVL arc welding equipment)
- plastic workers (some artificial UVL used in drying and curing)

- police officers
- printers (some artificial UVL used in drying and curing)
- railway maintenance workers
- restaurant/kitchen staff (artificial germicidal UVL)
- road workers
- sailors
- ski instructors
- sportsmen and women
- surgeons, nurses, operating room personnel (germicidal UVL)
- surveyors
- unprotected bench workers (adjacent to arc welding)
- welders (especially those using Tungsten Inert Gas), and
- wood curers (some artificial UVL used in drying and curing).

Ultraviolet radiation can have the following deleterious effects on the skin.

Sunburn: Reddening of the skin, blistering, and later peeling of the skin.

Keratoses: (Also known as sunspots.) Dry, rough, stable spots on the skin. These are not cancers but indicate excessive exposure to UV radiation.

Skin cancer: Various forms in Australia. Most dangerous form is malignant melanoma, which appears as a spot or mole that changes colour and shape over months.

Eye damage: Ultraviolet radiation can cause damage to the retina of the eye and cataracts of the eye.

Preventive measures

There are a number of steps employers can take to reduce the exposure of outdoor workers to sunlight. One essential element amongst these is the education of workers to the dangers that excess exposure to the sun can cause.

Other measures include:

- rescheduling work so that workers are inside between the hours of 10.00 am and 2.00 pm (11.00 am to 3.00 pm during daylight saving) when UV radiation is strongest
- taking advantage of existing shade or creating shade using canopies and other portable shade structures, and
- supplying and recommending personal protection devices such as hats with wide brims or long flaps, sunscreens, sunglasses and loose protective clothing to cover skin.

Other references

Guidance Note on the Protection of Workers from Ultraviolet Radiation in Sunlight, National Occupational Health and Safety Commission, 1991.

Protection from Sunlight, National Occupational Health and Safety Commission, 1992.

Skin Cancer and Outdoor Workers — a Guide for Workers, WorkCover NSW.

Skin Cancer and Outdoor Workers — a Guide for Employers, WorkCover NSW.

AS/NZS 4399: *Sun protective clothing — Evaluation and classification.*

See also the websites of government OHS authorities and other OHS information providers such as those listed in the Appendices.

OCCUPATIONAL SKIN DISEASES

¶10-420 Skin diseases

Adverse effects on workers' skin can result from exposure to a wide variety of substances and other influences at work.

"Dermatitis" is a general term used to describe any inflammation of the skin. "Contact dermatitis" refers to those conditions that result from direct contact of the skin with an external agent. "Irritant contact dermatitis" is caused by an irritant producing cell damage in the skin with which it comes in contact. Only the skin contacted by the irritating agent becomes affected. "Allergic contact dermatitis", which is less common, is caused when contact with a chemical agent provokes an allergic reaction.

Dermatitis that arises from within the body is known as endogenous. Endogenous skin diseases are rarely caused by occupational factors, although they may be aggravated by agents at the workplace.

Contact dermatitis is characterised by redness, swelling and blistering of the skin, then formation of scales and crusts. After continued irritation the skin becomes thickened and leathery, with spotty areas of pigmentation.

Other skin diseases associated with workplaces

This commentary will concentrate on the various forms of dermatitis, the major occupational skin disease. However, other skin diseases that may occur are described briefly below.

- *Skin keratoses and cancers*: skin damage caused by excessive exposure to the sun (in particular, to UV radiation). It is vital for Australian workers, who spend a large amount of time outside, to protect their skin against sun damage. This topic is considered in detail in the section on occupational cancers, at ¶10-360.
- *Contact urticaria ("hives")*: a reaction occurring within a few minutes to a couple of hours after exposure of the skin to certain substances. Balsams, alcohols, some antibiotics, nickel, platinum salts and cobalt are some examples of agents which can cause hives.
- *Photosensitisation*: some substances make the skin particularly vulnerable to sunburn. Examples include coal tar pitch and creosote (used as a wood preservative).
- *Occupational acne*: acne can be caused by the irritant effect of mineral oils on hair follicles in the skin. Workers on metal-cutting machines are

susceptible. Acne can also be associated with fats and oils used to prepare fried foods.

- *Chloracne*: a far more serious acne-like condition caused by contact with certain chemicals. The condition lasts for many years after exposure has ceased. Some dioxins (notably 2,3,7,8 tetrachloro-di-benzodioxin (TCDD)) are known to cause chloracne.
- *Scleroderma*: roughening and thickening of the skin. Agents known to induce scleroderma include silica dust, vinyl chloride monomer, trichloroethylene and tetrachloroethylene (perchloroethylene).
- *Folliculitis*: this refers to sores arising from irritation of hair follicles. Cutting oil used in association with metal-cutting equipment is the major cause.
- *Low humidity occupational dermatitis*: it is now recognised that particularly dry air can lead to rashes and other skin irritations. This is caused by the outer layer of the skin losing part of its water content. Affected skin also becomes more susceptible to irritants and allergens, that can cause contact dermatitis. Dry air is found in many modern indoor working environments. Those who work with furnaces and ovens and those who work in computer rooms may be particularly susceptible.
- *Perionychia*: this refers to inflammation around the fingernails, found in workers who frequently have wet hands. Bar workers, for example, are susceptible.

¶10-430 Common causes of occupational dermatitis

The following list sets out some of the major agents known to cause occupational dermatitis, along with the industries in which those agents are found. The list includes the most common causes, but is not exhaustive.

- cement — construction workers, cement manufacture workers, cement products
- epoxy resins — construction workers, applicators in light industry
- acrylic plastics — moulded articles producers, sealant users, printers
- surfactants, soaps, detergents — kitchen and hotel workers, hospital workers, metal industry
- mineral oils — metal machining workers
- solvents — degreasing and cleaning workers
- tar, pitch — roofing and bitumen workers, and handlers of creosote-treated wood, eg railway sleepers, electricity poles
- soldering fluxes — electronics workers
- pesticides and preservatives — manufacturing workers and applicators in farms, forestry, and users of sprayed products
- printing inks and plates — printing workers
- pharmaceuticals — manufacturing workers
- hairdressing preparations — salon workers, and

- film processing chemicals — film laboratory workers.

Cement is the single biggest cause of occupational dermatitis. The alkalinity of cement causes irritant contact dermatitis, while the chromate content can cause allergic contact dermatitis. It is recommended that workers handling cement, particularly wet cement, be provided with protective clothing to avoid skin contact. Ferrous sulphate can be added to cement products to prevent chromate dermatitis, although it is better if this is done by purchasing cement additives which already contain ferrous sulphate.

¶10-440 Diagnosis and treatment of dermatitis

Effective treatment of occupational dermatitis depends primarily upon an accurate diagnosis. A correct diagnosis can lead to the elimination/control of the noxious agent and thus to the prevention of similar problems in both the patient and other workers. At the same time, the appropriate treatment can be provided.

Because of the importance of correct diagnosis, all patients with suspected occupational dermatitis should be referred to a dermatologist. Even experienced clinicians may find it difficult to distinguish between dermatitis and other conditions, such as tinea.

The taking of a thorough history is essential to the correct diagnosis of dermatitis. This should include exposure to any known or suspected irritants or allergens. The possibility of irritants/allergens away from the workplace should also be considered. Any improvement or lack of improvement on weekends or holidays can be an important clue as to the existence of irritants/allergens at the workplace.

Suspected allergic contact dermatitis can be diagnosed through patch testing (note that patch testing is not appropriate for irritant contact dermatitis). Patch testing involves applying small amounts of suspected chemicals to the skin for 48 hours. There is a standard patch test battery, containing the most common contact allergens, as well as special batteries available for certain industries. Alternatively the worker may supply suspected chemicals or materials from the workplace for use in a patch test. Patch testing should only be performed and interpreted by a dermatologist.

Once a correct diagnosis has been made, removal from the causative agent and the appropriate treatment should result in cure. Employers may need to take the following steps to remove the affected employee from contact with the irritant/allergen:

- provide protective clothing
- move the affected worker to a new area within the workplace, or
- replace the offending agent with an alternative (for example, a particular brand of printing ink containing a known allergen may be able to be replaced with another brand free from the allergen).

In a small number of cases the only way for a worker to avoid the noxious agent will be to change his/her profession.

¶10-450 Preventive measures

In its *Guide to Occupational Diseases of the Skin*, the NOHSC recommends a number of measures that can help to prevent the occurrence of occupational skin diseases. These are summarised below.

1. Where possible, the hazard should be eliminated. If this is not possible, a less hazardous chemical or process should be substituted.
2. Engineering controls should be used to minimise exposure to dust, gases, fumes, fogs, mists and vapours. These controls should aim to prevent direct contact of the dust or solution with the skin, as well as avoiding build up of materials on surfaces that may come into contact with the skin.
3. A high standard of hygiene should be maintained in the work environment.
4. A high standard of personal hygiene should also be applied. If necessary, employees should have access to clean work clothes, protective clothing, laundry facilities, showers and change-rooms.
5. Employees should be trained in the correct use and care of protective clothing and equipment.
6. Material safety data sheets should be consulted for details of skin hazards in relation to new and existing compounds.

Further references

Health and Safety Bulletin No 49 and 50, "Guidelines on Occupational Skin Diseases", May and June 1987, ACTU/VTHC Occupational Health and Safety Unit.

Cement Dermatitis, National Occupational Health and Safety Commission, 1993.

Occupational Diseases of the Skin, Guide, November 1990, National Occupational Health and Safety Commission.

Adams, RM (ed), *Occupational Skin Diseases*, 2nd edn, WB Saunders, Philadelphia, 1990.

The Journal of Occupational Health and Safety — Australia and New Zealand, Vol 2, No 4, August 1986 (special issue on occupational skin diseases) (CCH Australia Limited).

See also the websites of government OHS authorities and other OHS information providers such as those listed in the Appendices.

WORKPLACE NOISE AND HEARING LOSS

¶10-520 Extent of the problem

Noisy working environments and the use of noisy equipment can, over a period of years, cause significant loss of hearing. Occupational hearing loss, often referred to as noise-induced hearing loss or "industrial deafness", results from irreparable damage to the delicate structures in the inner ear. The effects are likely to be gradual and, coupled with loss of hearing ability due to advancing age, can be substantial.

It is estimated that between 250,000 and 500,000 Australian employees work in hazardous noise environments.

While leisure activities can also contribute to hearing loss, the extent of this can be overemphasised. Away from the workplace, exposure to loud noise tends to be infrequent and for short periods, though listening to loud music, loud audio systems in motor vehicles and frequent attendance at venues where very loud music is played can damage hearing. Workers at such venues face very real risks to their hearing.

The consequences of hearing loss can be extremely disturbing to the individual. The reduction in communication ability is the most serious, although there are all sorts of other domestic noises which we usually take for granted (television and radio, hearing the kettle boil, a car horn signalling warning, and so on). Noise-induced hearing loss often causes people to become resentful and places additional strains on relationships due to the problems in communication. There may be a marked reluctance for a person to admit that they have suffered some hearing loss, and consequently to seek assistance.

At work a noise-induced hearing loss can lead to an inability to hear instructions clearly with consequent poor relationships with fellow workers and immediate supervisors. It may lead to misinterpretation of instructions, and could possibly contribute to accidents and loss of continuity in the work. Telephone conversations against the background noise of a workshop become virtually impossible.

¶10-530 Measurement of hearing loss

A possible noise hazard and the associated hearing loss is indicated informally by such factors as:

- difficulty in communication by speech at the workplace (having to raise the voice at a distance of one metre or less)
- ringing in the ears (tinnitus) or difficulty in hearing after exposure to noise
- temporary loss of hearing or shouted conversation at the end of the workday, and
- an unusual prevalence of deafness in employees.

Informal measurement, however, is not conclusive. A noise survey (involving measurement of noise levels) should be undertaken in all areas where hazardous exposure to noise is known or suspected to occur, because an individual will usually only become aware of a change in hearing after damage has occurred. The measurement of noise levels is discussed further at ¶10-540.

The measurement of a person's hearing ability is known as audiometry. It should only be carried out by a person trained in that skill. An audiometric test generally involves sitting in an insulated room and listening on headphones to sounds of differing tone and loudness. While it is possible to distinguish between noise-induced hearing loss and hearing loss caused by disease, it is not possible to distinguish between noise-induced hearing loss and loss brought on by old age. Nor is it possible to determine which noises were responsible for any loss.

Where the workplace involves a reasonable amount of noise, it is wise to test the hearing acuity of all new employees before they start work. This is:

- for their own information
- to ensure that in any future compensation claim the employer will not pay for a hearing loss that was not caused at that workplace, and
- to create awareness of a workplace hearing conservation program.

After this, tests could be carried out every two years. Note that in some states there are legislative requirements prescribing regular audiometric testing.

¶10-540 Measurement of noise levels

The unit of noise is the decibel. A modified weighting scale measured over the range of audible frequencies has been developed, known as the "A scale". This is an attempt to assess the intensity and frequency of the sound so that it is similar to the way a normal healthy ear hears the sound. Its units of measurement are expressed as dB(A). Decibels are measured on a logarithmic scale, so the 80 dB(A) is 10 times 70 dB(A) and 90 dB(A) is 100 times 70 dB(A). Sound levels for some common noises are provided below.

Sound level in dB(A)	Sound source
140	Jet engine
130......................................	**THRESHOLD OF PAIN**
120	Propeller plane
110	Chain saw
100	Sheet-metal workshop
90	Heavy truck
80	Heavily trafficked street
70	Car
60	Normal conversation
50	Low conversation
40	Quiet radio music
30	Whispering
20	Quiet urban apartment
10	Rustling leaves
0..	**HEARING THRESHOLD**

Because the dB(A) system is measured along a logarithmic scale, exposure should be halved for every 3 dB(A) increase. This means that while exposure to noise levels at 85 dB(A) for eight hours may be tolerable, exposure to noise at 88 dB(A) should be limited to four hours, exposure to 91 dB(A) should not last longer than two hours, and so on. In general, the legal requirement throughout Australia is to limit workers' noise exposure to a maximum of 85 dB(A) for eight hours, or equivalent exposure.

Noise measuring equipment

Noise level meters are used to determine the level of noise at a particular place. Some instruments can measure sound levels at different frequencies.

¶10-550 Noise management programs

The basic steps for noise management and conserving hearing are set out in Australian Standard AS/NZS 1269 Set:2005 *Occupational Noise Management*. The process involves the measurement of noise, its reduction where feasible, the use of hearing protection if reduction is not feasible, and audiometry to monitor overall progress. Further essential ingredients include the motivation of workers, allocation of specific responsibilities to supervising staff and involvement of those concerned with safety and/or health in the workplace.

Engineering controls

The removal of noise by engineering methods should be given top priority in any hearing conservation program. Noise specifications should be included in a company's purchasing criteria, and all new equipment should be checked before purchase for the noise that it will produce. Installation of machinery should also be done with noise control in mind.

There is a wide range of steps that can be taken to reduce the noise caused by machinery that is already in place. These include:

- replacing noisy machines with quieter ones
- placing machinery in sound-proofed rooms
- placing machinery on vibration-dampening mounts
- using exhaust-dampening equipment, and
- treating walls to reduce reflected noise.

Administrative controls

Some administrative measures may be introduced to minimise workers' exposure to noise levels. These could include the rescheduling of noisy processes to periods of minimal occupancy, job rotation and the signposting of noisy areas. Often administrative measures can be used as a temporary solution until engineering solutions can be implemented.

Personal hearing protection

In some cases, elimination of harmful noise is not feasible so that hearing protection will be necessary for the foreseeable future. Fortunately, the

Australian method for testing hearing protectors and for rating their performance with a single number has received wide acceptance because it is conservative and relates closely to how such items are used in practice. However, personal protection is never a really satisfactory method of risk control, for reasons such as the following:

- it is not possible to ensure that hearing protectors fit all the time, so they cannot be regarded as giving full protection all the time
- hearing protection may be uncomfortable to wear, particularly in hot and humid weather
- ear-muffs become damaged and worn — a system of regular maintenance must be introduced, and
- ear infections are possible through the fitting of plugs when proper personal hygiene is not observed.

Introducing a program

A hearing conservation program should be introduced with the full support of management, employees and technically competent persons. The cooperation of employees should be sought in suggesting and designing control measures. An education program can accompany the introduction of the noise conservation program. This should be aimed at all employees, and should cover the effects of noise, its measurement and the use of hearing devices.

The fact that each year sees quieter equipment coming onto the market and that hearing protectors compatible with a wide range of other personal protective equipment are now available negates any excuse for delaying the introduction of hearing conservation programs in the majority of small organisations, as well as in large industries.

¶10-560 Legislative provisions and standards

Noise levels and hearing conservation are covered by regulations in each state and territory. The exposure standards that have been adopted in all jurisdictions are:

- an eight-hour equivalent continuous sound pressure level (L aeq.8h) of 85 dB(A), referenced to 20 micropascals, and
- a peak sound pressure level (L peak) of 140 dB(C).

National standards

The NOHSC has established a national standard for exposure to occupational noise of an eight-hour noise equivalent of 85 dB(A), with a maximum peak noise level of 140 dB(C). All states and territories have adopted this exposure standard in legislation.

AS/NZS 1269 Set:2005 *Occupational Noise Management* includes five standards which together provide an integrated approach to establishing, implementing and evaluating an occupational noise management program. Its aim is to help organisations reduce the effects of noise at work.

Regulations and codes of practice

NSW

Occupational Health and Safety Regulation 2001

Code of Practice: Noise Management and Protection of Hearing at Work

Vic

Occupational Health and Safety (Noise) Regulations 2004

Code of Practice for Noise

Qld

Workplace Health and Safety Regulation 1997

Code of Practice for Noise Management at Work

SA

Occupational Health, Safety and Welfare Regulations 1995

WA

Occupational Health and Safety Regulations 1996

Code of Practice for Noise Control in the Workplace

Control of Noise in the Music Entertainment Industry

Tas

Workplace Health and Safety Regulations 1998

NT

Work Health (Occupational Health and Safety) Regulations 1992

Adopted national code

ACT

Adopted national standard and code

Further references

Noise Problems at Work — Guide for Assessing and Fixing, WorkSafe Victoria, May 2005.

Code of Practice: The Control of Noise in the Music Entertainment Industry, WorkSafe WA Commission, 2003.

Noise Management at Work Resource Package, a kit containing a control guide publication, a video, fact sheets, a poster and stickers, produced by the NOHSC.

National Standard for Occupational Noise, [NOHSC:1007(2000)], National Occupational Health and Safety Commission.

National Code of Practice for Noise Management and Protection of Hearing at Work, [NOHSC:2009(2004)], National Occupational Health and Safety Commission.

Control Guide: Noise Management at Work , National Occupational Health and Safety Commission, 1991.

WorkSafe Western Australia has over 40 items on noise management on its website at www.safetyline.wa.gov.au.

Noise Management: a Trainer's Guide, ACCI, 1990.

Guidelines for Control of Workplace Noise, WorkCover Corporation of South Australia, Adelaide, 1998.

Noise at Work, WorkCover NSW.

Noise Regulation — Hospitality and Entertainment Industry, WorkCover NSW.

SHARE Manual of Documented Solutions to Noise, Health and Safety Organisation (Vic).

Noise at Work, Department of Occupational Health, Safety and Welfare (WA).

Managing Noise at Work, Department of Employment, Vocational Education, Training and Industrial Relations (Qld).

Occupational Noise, Northern Territory Work Health Authority.

See also the websites of government OHS authorities and other OHS information providers such as those listed in the Appendices.

OCCUPATIONAL OVERUSE SYNDROME

¶10-620 The nature of occupational overuse syndrome

Occupational overuse syndrome (OOS, also known as repetitive strain injury) is a term used to cover a variety of soft tissue injuries affecting various parts of the body. These conditions are attributed to various types of work and recreation where overload of soft tissues (such as muscles and tendons) occurs through rapid, repetitive and/or forceful movements, and/or where static muscle loading is maintained over a period of time. Injury is experienced as pain and weakness that may lead to impaired function of the muscles and other soft tissues.

Occupational overuse syndrome affects various parts of the body, including the neck, shoulders, elbow joint, hands, fingers and wrists. More rarely, people using their feet to operate machinery may suffer injuries in the lower parts of their legs. Even voice strain (''losing one's voice'', or hoarseness) can be a type of overuse injury in professions such as teaching, auctioneering or race-calling. Probably the best known type of overuse injury is tenosynovitis, which is a strain injury of a tendon or group of tendons about the wrist, causing inflammation of the tendon sheath. Its main symptoms are pain and swelling over the affected tendons, increasing restriction in the use of the hand, progressive weakening in grip strength and a creaking or crunching sensation felt in the affected area upon movement of the thumb.

Causes

The following factors can contribute towards injuries:

- awkward body postures, including postures workers are forced to adopt by poor design of equipment or work layout
- poorly designed workstations, equipment, machinery and tools not matched to the employee, including the effects of vibration and sudden impact forces
- poorly designed tasks, that is, factors such as employee position, forces required and the design and placement of equipment
- work organisation factors which may contribute to demands placed on employees, such as required output, duration and variation of tasks, number and duration of pauses and the urgency of deadlines
- inappropriate/poor arrangement of job design, for example, the requirement to perform the same repetitive movements, and
- new employees, or those returning to work after an extended absence, being required to perform repetitive movements without a period for adjustment.

Work organisation can also be responsible to some degree for the onset of OOS. Boredom, shiftwork, poor communication and lack of control over a job may all contribute. Stress may also be a factor. Studies indicate that a greater variety of work, regulating work flow to avoid rush periods, and job enrichment can help reduce the level of OOS (see ¶10-760 on reducing the causes of stress).

The syndrome is not only caused by work in a work setting. The combination of the type of work performed and the nature of recreational activities or domestic tasks may be a causative factor for some individuals. Some individuals may be predisposed to the development of this condition, although this theory has not yet been proven.

¶10-630 Types of workers affected

As research into this area broadens, the range of occupations considered vulnerable to occupational overuse injuries has widened. Whereas initial studies concentrated mainly upon factory and manual work, a number of white-collar/clerical areas have now been identified as prone.

The NOHSC code of practice provides the following occupations where occupational overuse injuries have been noted to occur:

- process workers
- machinists
- cleaners
- kitchen workers
- keyboard operators
- clerks
- musicians

- carpet layers
- hairdressers, and
- mail sorters.

This list is not exhaustive.

¶10-640 Reducing the causes

The NOHSC's *National Code of Practice for the Prevention of Occupational Overuse Syndrome* has been adopted by most states and territories. Victoria has developed its own code of practice. The national code states that to:

> "identify and address the causes of OOS it is necessary to consider many aspects of the workplace. The overall conditions of work, including the industrial relations climate, will influence the effectiveness of changes introduced to prevent OOS."

Elimination of possible causes of OOS will involve attention to job design, work layout and ergonomic aspects such as the following:

- analysis of equipment, tools and working posture to reduce physically stressful movements and postures and allow a variety of movements and postures within each task to rest muscle groups
- dimensions and positioning of aspects of work layout in relation to the worker's strength and size
- equipment design, such as handles, grip sizes, provision for adjustment, suspension of heavy items
- training and instruction of employees (both when commencing and resuming work after suffering overuse injury) in recommended work methods, efficient use of muscles and gradual introduction to repetitive manual tasks
- job rotation so that workers perform different tasks, and
- regular rest breaks from constant and rapid repetitive tasks, as well as encouraging workers to vary the rhythm of hand movements when performing these tasks. In some cases, where it is not possible for keyboard operators to perform other tasks, agreements have been negotiated that provide for work pauses of 10–15 minutes in each hour. The introduction of pause gymnastics has been useful in some workplaces.

Proper education of workers subject to occupational overuse injuries, and their supervisors, is also an important part of prevention. Education should ensure that employees and supervisors understand the potential health problems associated with their workplace and how to avoid those problems. For example, keyboard operators should be informed about the correct use of furniture and equipment and the importance of correct posture, movement and exercise. Employees may also be informed of the relevance of domestic/ leisure activities and their potential for worsening OOS.

Occupational overuse syndrome should not be seen as a purely medical or mechanistic phenomenon. In its consideration of the syndrome, the Australasian Faculty of Occupational Medicine concluded that "it seems many claims of RSI [as OOS was known in the 1980s] were in fact people's communications of discontent or alienation in their employment. Seen in this light, it is not surprising that casting the problem in medical terms and advising people to "see your doctor if pain persists" did not produce the solution. Better results were achieved when managers were forced to confront the issue and properly address workers' complaints."

Overall, therefore, reducing the causes of OOS may not be a straightforward matter, and may require the advice and assistance of experts in the field, as may the rehabilitation of injured workers. There is no point in embarking on partial interventions such as only buying ergonomic furniture. Causes are multiple and the solution is complex. Partial attempts at rectification/correction will lead to an unsatisfactory result.

Further references

National Code of Practice for the Prevention of Occupational Overuse Syndrome, [NOHSC:2013(1994)], National Occupational Health and Safety Commission.

Guidance Note for the Prevention of Occupational Overuse Syndrome in Keyboard Employment, National Occupational Health and Safety Commission, January 1989.

Guidance Note for the Prevention of Occupational Overuse Syndrome in Manufacturing Industry, National Occupational Health and Safety Commission, 1992.

Gore, A and Tasker, D, *Pause Gymnastics: Improving Comfort and Health at Work*. A book of exercise programs designed for performance during short work breaks. The exercises are specifically devised to prevent, reduce and eliminate discomfort and pain associated with musculoskeletal injuries such as OOS (CCH Australia Limited).

RSI Retrospective — Consensus Statement, Australasian Faculty of Occupational Medicine, 1995.

See also the websites of government OHS authorities and other OHS information providers such as those listed in the Appendices.

STRESS

¶10-720 Rise in stress claims

A steady increase in the number and cost of claims for work stress can be seen in workers compensation statistics over the last decade. For example, in New South Wales, stress-related problems accounted for more than one in three occupational disease claims in 2002–2003; while in Victoria, in the 10 years from 1992, stress claims rose from 3.6% to 8% of all workers compensation claims. More striking is the difference in cost, with stress claims typically costing much more than claims that are not stress related, because they usually involve longer periods away from work.

What is stress?

In simple terms, stress could be defined in the context of occupational health as the result of the body's reaction to excessive levels of pressure caused by various environmental factors. The level of reaction is strongly influenced by the person's own perception of the stress-causing situation, this perception in turn being a function of personality. Stress could therefore be seen as a set of physical, mental and biochemical conditions within a person's body that reflect the body's attempt to adjust to pressuring situations. In the work context, pressuring situations are those aspects of a job that cause problems for the employee and that require some form of adaptation to their demands.

As these situations occur, certain mechanisms in the body are activated, such as glands releasing hormones or changes in the central nervous system, thus enabling the body to adapt. If exposure to the situations is prolonged, these mechanisms will wear down in time until they become inadequate, and symptoms of excessive stress will develop. It is the continuing nature of demands which leads to stress.

Thus, a stressful work environment can impair general health and wellbeing. It is recognised that there are certain conditions, such as alcoholism, unstable mental states and certain diseases (see ¶10-740), in which stress may be inferred as a component either allowing a vulnerability or exacerbating an existing state. Also, stress can be seen to give rise to injury more directly where, for example, safety precautions are ignored to meet a perceived "pressure" of deadlines or quotas.

This information implies that stress is always a bad thing, which is incorrect, as everyone has an optimal level of stress at which they perform most effectively. A degree of stress can be pleasant and beneficial, and it could be said that people require a certain amount of stress for their lives to be meaningful. The problems begin when the optimal level is exceeded. (Because of these difficulties with the stress concept, some researchers take "stress" as being a term to simply describe a body's adaptive response, then differentiate between "distress" where the adaptive response is perceived to be harmful and "eustress" as in euphoria where the response is at a level of pleasant stimulation.)

Early studies of stress concentrated upon its effects at management and executive levels, but it is now apparent that stress can occur at all employee levels within an organisation.

¶10-730 Causes of stress

The range of potential stress-causing situations to be considered at work is very wide, but can be divided into three basic categories, as shown in the following table:

1. Job-induced	
• work overload/information overload • role conflict • lack of control over work environment and decision-making (affects both blue and white-collar workers) • excessive work pace or deadlines	• work underload • role ambiguity • unsatisfactory feedback • insecurity • continual concentration upon goals; overcompetitive work environment
2. Organisation-induced	
• working conditions (noise, light, poor ergonomics, lack of privacy, vibration, overcrowding) • office politics and idle talk • inappropriate management styles • unsafe conditions • unsatisfactory work equipment • various communication problems	• lack of management support • lack of peer group support • discriminatory practices • sexual harassment • unsuitable staffing levels • lack of training • bullying
3. Change-induced	
• unfamiliarity with new job • expectations of change, such as transfer, retirement, retrenchment	• changes in technology and work systems • changes in plans, procedures and work layout • new supervisor • new co-workers/subordinates

In addition, stress can be induced by domestic and personal problems which are "brought to work" and may interfere with the employee's ability to work satisfactorily by causing conflict of needs and priorities.

Note also that the causes of stress can flow downwards from superiors, upwards from subordinates or laterally from peers (such as office politics).

¶10-740 Effects of stress

Employee stress can have drastic effects on performance, productivity and cost levels because the affected employees are prevented from efficiently carrying out their work. Apart from the unfortunate effects on the individual, stress can cause adverse reactions throughout a workplace, such as effects upon employee relations, workgroup attitudes and disruption to the general work flow.

For the individual, there are three main types of response to stress:

1. psychological changes
2. behavioural changes, and
3. forms of disease or illness.

Psychological changes that may be induced by stress include feelings of boredom, guilt, pressure, anxiety, tension, irritation, worry and pessimism.

Among the many types of behavioural changes that may provide "clues" to stress are absenteeism, cynicism, deteriorating work performance, "addiction" to work, loss of interest and effort in work, withdrawal, conflict/abrasiveness, and excessive use of coffee, nicotine, alcohol or medicines.

Stress is able to cause physiological changes in the body, and it is possible to measure such changes by measuring the release of certain hormones. For instance, acute stress may lead to the release of cortisol, adrenaline and noradrenaline from the adrenal glands, which can be measured to determine whether they are above usual levels. These physical changes may be associated with various illnesses, including coronary/heart diseases, hypertension, ulcers, migraines, psychosis, neurosis, depression, skin disorders, insomnia and diabetes.

¶10-750 Stress and the individual

An assessment of the causes and effects of employee stress should bear in mind that individual responses to stress-inducing situations will vary widely. Some people are able to tolerate a great deal more stress than others, because they do not react as adversely towards it. Also, the same individual's response will vary over time so that he/she may be more or less vulnerable to the adverse effects of stress.

The difference between individuals can be attributed to many factors: personality type, age, education, ambitions/expectations, social background, marital status, level of health/fitness before commencing a job, tendency to smoke, consume alcohol and so on. The effect of factors such as these should be considered when examining the methodology of various studies of stress at work. They can be regarded as factors which are brought to the job.

While it is true that certain individuals will be more susceptible to the effects of stress, this should not be used as an excuse for failing to address workplace stressors. All workers will respond positively to the alleviation of an overly stressful situation.

¶10-760 Methods of alleviating stress

Role of the employer

The main responsibility of the employer is to minimise the opportunities for stress to develop. This affects several aspects of human resources management — job design, job satisfaction, working conditions, participation in decisions, clear-cut and equitable policies, training and selection of employees, adequate pay rates, grievance management procedures and so on.

Training people to deal with stress

From an individual viewpoint, employees may be able to cope with stress better if they can make use of techniques to relax or seek some other means of coping with the situation. It may therefore be beneficial for management to provide some form of "stress education" to employees, in forms such as seminars, videos, printed information or individual counselling.

Stress management courses may help, but by far the best approach is to remove or reduce as many of the causes of stress at the workplace as possible. When introducing such a course, management should demonstrate that they are not trying to place the responsibility for stress solely on the employees.

Treating stress

There are several ways of reducing stress at the workplace. They can be divided into the following broad categories.

1. Attention to the *work environment* (such as job redesign and autonomous workgroups).
2. Improving the *interpersonal environment* (for example, by meetings, more positive feedback, team-building approaches and better communication).
3. Assisting the *person* — counselling, psychotherapy, rehabilitation, psychological techniques (such as transcendental meditation, transactional analysis, biofeedback, support groups) and assistance with maintaining health and physical fitness (such as providing gymnasium facilities or encouraging health awareness).

Improving communication, especially in terms of allowing employees a say in decision making, is also likely to help by reducing uncertainty and fears. For example, the introduction of new technology to a workplace without discussion and adequate preparation may cause fears of retrenchment. Adequate communication would remove this particular stressor.

Further references

Brecht, G, *Stress at Work*, Proceedings of the Fifth National Conference of the Employee Assistance Professional Association of Australia, April 1997.

How to Manage Staff and Stress Responses, Comcare Australia, Canberra, 1992.

"Critical Factors in Understanding Workplace Stress: the Theory and Reality" in *Reducing the High Cost of Stress and Absenteeism*, Comcare Australia, Conference Proceedings 24–25 August 1995, Sydney, NSW, IES Conferences, 1995.

McCarthy, P, Sheehan, M and Wilkie, W, *Bullying from Backyard to Boardroom*. Sydney, NSW, Millenium, 1996.

Morehead, A et al, *Changes at Work: the 1995 Australian Workplace Industrial Relations Survey*, Canberra, ACT, AGPS, 1997.

Stress at Work — Not What We Bargained For, ACTU, 2003.

Stress, the Workplace and the Individual, WorkCover NSW, 1996.

See also the websites of government OHS authorities and other OHS information providers such as those listed in the Appendices.

SMOKING AND THE WORKPLACE

¶10-820 Health effects of exposure to smoke

Exposure to workplace tobacco smoke is an area where requirements have been progressively tightened since the early 1990s. Changes in employers' legal obligations reflect raised community awareness of the health damage which can result from exposure to other people's cigarette smoke at work.

This is known as "passive smoking", and given the amount of time spent at work, it can have serious health consequences.

Smoking is now banned in most public sector workplaces, while smoking bans or restrictions are becoming increasingly common in the private sector. In 1990 the NOHSC published the *National Policy Statement on Smoking and the Workplace*, that states that all Australian workplaces should aim to become smoke-free. Note that smoking is now banned in many areas of public life (public transport, cinemas, domestic flights).

Tobacco smoke contains thousands of potentially harmful chemicals, including:

Cancer-causing agents (carcinogens). There are at least 43 known carcinogens in tobacco smoke. One of the most potent of all carcinogens is benzo-pyrene, discovered in tobacco smoke over 30 years ago.

Carbon monoxide. This is a poisonous gas that lowers the amount of oxygen carried by the blood.

Nicotine. This is the addictive drug that maintains the tobacco habit.

Radioactive compounds. The radioactive compounds found in highest concentration in cigarette smoke are polonium 210 and potassium 40. Radioactive compounds are known to cause cancer.

Hydrogen cyanide. In the amounts found in tobacco smoke this gas kills cilia, the tiny hairs that move together in waves to help keep our lungs clean.

Pesticides. A range of pesticides have been found in tobacco, including DDT, endrin, parathion and endosulfan.

Metals. Many toxic metals, including arsenic and nickel, have been found in cigarette smoke.

Some tobacco smoke poisons are more concentrated in sidestream smoke (that smoke released directly into the air from the burning end of a cigarette) than in mainstream smoke (that smoke breathed in by the smoker). The concentration ratios are:

Poison	**Ratio of concentrations Sidestream: Mainstream**
Carbon monoxide	8:1
Nicotine	3:1
Formaldehyde (inhibits lung cilia)	51:1
Benzo-pyrene (carcinogen)	3:1
B-napthylamine (carcinogen)	39:1
4-Aminobiphenyl (carcinogen)	30:1
Dimethyl nitrosamine (carcinogen)	52:1

However, the concentrations of some of these substances in the air where people are smoking may be so low that the health hazard they present is negligible. The degree of risk depends on the degree of exposure.

Lung cancers and the increased risk of heart disease are the most serious consequences of passive smoking. The non-smoking wives of smoking husbands have been shown as having a significantly greater risk of developing lung cancer than non-smoking wives of non-smoking men in the same population. Some other consequences of involuntary exposure to tobacco smoke include the possibility of:

- aggravation of the condition of persons suffering angina pectoris, the pain caused by diminished blood supply to the heart muscles
- precipitating attacks in asthma sufferers, and
- measurably impairing lung function in long-term passive smokers.

Eye irritation (discomfort, itching, watering), nose and throat irritation, hoarseness, wheezing and headaches are commonly reported annoyances associated with the presence of tobacco smoke. People with allergies are especially susceptible to these irritations, and up to 20% of the population are included in this category.

The links between passive smoking and adverse health effects are now well established. In February 1991 the Federal Court of Australia held that there was an established link between passive smoking and the development of lung cancer, asthma and respiratory diseases.

This was confirmed in 1992 in the New South Wales District Court, when a jury found that passive smoking can be dangerous to health. It awarded $85,000 to a psychologist employed by the New South Wales Department of Health. She claimed that her employer had been negligent in exposing her to other people's cigarette smoke at her workplace. The smoke affected her pre-existing asthma and lung disease.

In another case heard before the New South Wales Supreme Court in 2001, a woman who claimed that her workplace exposure to environmental tobacco smoke resulted in throat cancer was awarded $450,000 in damages. Her employer was found to be negligent in failing to provide a safe workplace.

Legislative prohibitions on smoking

In October 2003, the National Occupational Health and Safety Commission updated its position statement against smoking in the workplace. In its *Guidance Note on the Elimination of Environmental Tobacco Smoke in the Workplace*, the NOHSC called for workplaces to be made "completely smoke free as soon as possible".

The guidance note claimed that designated indoor smoking areas within workplaces are not considered effective controls, and maintained that mechanical dilution ventilation does not adequately eliminate exposure to environmental tobacco smoke (ETS) for any given level of smoking. Thus, the NOHSC does not endorse the use of Australian Standard AS 1668.2-2002 *The use of ventilation and air-conditioning in buildings, part 2: Ventilation design for indoor air contaminant control*. The guidance note recommended that smoking be restricted to areas that have full open air natural ventilation, and places where ETS cannot drift or be drawn into non-smoking areas.

The NOHSC guidance does not have the force of law. It is up to the states and territories to enact legislation to ban smoking in the workplace.

In most states and territories there are specific legislative provisions which prohibit smoking under certain circumstances, for example in lead process areas, or in the vicinity of asbestos work or certain dangerous goods.

There are certain workplaces where passive smoking may be particularly dangerous. Exposure to asbestos, radon, arsenic, aromatic amines and crystalline silica can interact with tobacco smoke to greatly increase the risk of lung cancer.

In addition, laws in all states and territories are progressively introducing restrictions on smoking in enclosed public places.

In New South Wales, for example, smoking in all "enclosed public places" (except for certain designated "exempt areas") was banned from 6 September 2000 under the *Smoke-free Environment Act 2000*. The definition of "enclosed public places" includes many different kinds of workplaces, such as:

- shopping centres, malls and plazas
- restaurants, cafes, cafeterias, dining areas and other eating places
- childcare facilities, schools, colleges and universities
- professional, trade, commercial and other business premises
- community centres or halls and places of public worship
- theatres, cinemas, libraries and galleries
- fitness centres, bowling alleys and other sporting and recreational facilities
- trains, buses, trams, aeroplanes, taxis and hire cars, and ferries and other vessels
- hostels, motels, hotels and clubs (other than residential accommodation)
- nightclubs and casinos (other than a casino private gaming area or residential accommodation)
- hospitals and nursing homes.

The *Smoke-free Environment Amendment Act 2004*, which commenced on 3 January 2005, removes certain exemptions from the 2000 Act and sets temporary exemptions for areas of a club, hotel, nightclub or casino. From 4 July 2005 to 1 July 2007, temporary exemptions will be allowed for one room, or part of room, in a club, hotel, nightclub or casino to be set aside as the exempt area. The exempt area must not exceed 50% of the total area of the rooms from 4 July 2005 to 2 July 2006; or 25% from 3 July 2006 to 1 July 2007. If the total area does not exceed 100 m^2, the exempt area must not exceed 50% from 4 July 2005 to 1 July 2007. The minister may authorise a club, hotel or nightclub to set aside a second room, or part of one, as part of the exempt area. Dining areas and certain other specified areas may not be set aside as exempt areas.

Other states and territories are introducing similar requirements. Local legislation should be checked for the details of requirements.

Other issues

Non-smokers' objections may also be grounded in non-legal, non-health issues, such as assertions regarding unpleasant odours in the workplace or in clothing, unsightliness and so on. Claims have also been made regarding the productivity of smokers — their absenteeism rates are higher and they take more breaks per day. It has also been asserted that smokers create additional costs in terms of cleaning and maintenance, property damage and depreciation.

Finally, there is the potential for friction between employees regarding the relative rights of smokers and non-smokers.

¶10-830 Risk management and smoking

There is an obligation on the employer to provide a safe and healthy working environment, so a workplace where employees are not exposed to ETS should be the goal of every employer.

Consultation with employees helps to ensure that the no-smoking policy is accepted and adhered to. The aim of consultation should be not to ask employees whether they want a non-smoking workplace, but to convey information about the health risks of exposure to ETS and the employer's obligations to protect people from foreseeable risks to their health, and to seek their cooperation and engagement in this endeavour. The workplace health and safety committee may be an appropriate forum through which the consultation can be conducted.

Policies banning smoking

The most effective way to stop occupational health problems associated with smoking is to ban smoking from the workplace. However, this may encounter some resistance if not implemented properly.

A smoking ban should be announced well in advance (at least three months) of the time it is to take effect. That time should be used to educate employees on the dangers of smoking (both passive and active) and to help them adjust. It may be possible to conduct an in-house quit program during that time, or to offer smokers paid leave to attend such a program.

The ban must extend to all levels of employees. A smoking ban that does not apply to executives would have little chance of acceptance.

Further assistance

Cancer Councils, the National Heart Foundation and Health Departments in the various states may offer further assistance. Various private organisations offer smoking cessation courses designed for workplace implementation. The New South Wales Department of Health has produced suggested guidelines for a smoking at work policy in its publication *Towards a Non-smoking Policy in the Workplace*.

Further references

The Health Effects of Passive Smoking: a Scientific Information Paper, National Health & Medical Research Council, Canberra, 1997.

Guidance Note on the Elimination of Environmental Tobacco Smoke in the Workplace, [NOHSC: 3019(2003)], National Occupational Health and Safety Commission, 2003.

Going Smoke Free — a Guide for Workplaces, National Heart Foundation, 1991.

Passive Smoking in the Workplace: Policy and Control — a Guide to Achieving a Smoke-free Workplace, WorkCover NSW, NSW Cancer Council, Sydney, 1997.

Passive Smoking in the Workplace, WorkCover NSW, 1997.

Achieving Smoke Free Workplaces, Division of Workplace Health and Safety (Qld).

Smoke Free Workplaces Code of Practice (ACT).

See also the websites of government OHS authorities and other OHS information providers such as those listed in the Appendices.

Chapter 11

MANAGING SOCIAL ISSUES

SHIFTWORK AND EXTENDED WORKING HOURS

¶11-120 Shiftwork and health

Health, safety and social issues are often connected, and their interrelatedness is clearly evident in discussions of shiftwork, night work and extended working hours. The interaction between work schedules and physiological and social functioning can have damaging effects on individuals, and risks in this area need to be managed as well as any other health or safety risks.

Types of shiftwork

Shiftwork may take the form of a regular, fixed schedule (such as a night watchman working the same shift every night) or may occur as part of a rotating shift system. It may also occur when people work a number of casual jobs in order to enhance their income.

In rotating shift systems, the most common periods of rotation are long periods (20 days or more) or weekly rotations. Some establishments are now moving to short rotations, with only two or three days on each shift. Twelve-hour shifts are becoming increasingly common, as they seem to involve less disruption to a shiftworker's life. Twelve-hour shifts allow for two days (or one) on day shift, then two days (or one) on night shift, followed by four days (or two) off.

Generally, night work offers better financial benefits than day work, and some people find it offers them social benefits as well.

Stand-by periods are another form of working outside usual hours. In some professions (notably maintenance/repair workers and emergency workers) employees may spend a large amount of their spare time "on call". The problems associated with stand-by periods are discussed at ¶11-150.

¶11-130 Effects on health

There is some controversy as to how serious the effects of continuous night work or shiftwork are. While some studies have claimed that significant adverse effects on employee health may occur over a period of time, others have concluded that there is little correlation between night/shiftwork and deterioration in health. Self-selection among shiftworkers and individual perceptions of a situation may be significant factors.

As with stress, the effects are likely to depend very much on the individual, but some guidelines can be drawn up to indicate the potential harmful effects of shiftwork or night work.

Night work involves a reversal of normal biological peaks and troughs (such as brain function, respiration) that are usually at their most active during the day, but are now called on for greater activity at night.

It is possible to test the body's adaptation to night work by measuring the level of certain hormones that are produced in greater quantities during the day. This research suggests that workers never really adapt to working night shifts. The main reason for this appears to be that shiftworkers invariably revert to "normal" sleeping patterns during weekends, breaks and holidays.

As a result of this, fatigue levels are increased by both the extra exertion needed to work during the "deactivation" period and having less chance to recuperate through sleeping during the "active" period of daytime. Sleep during the day tends to be less deep, more fragmented and shorter by an average of one to two hours. Thus shiftwork can lead to excessive fatigue, which in turn may cause exhaustion or psychosomatic illnesses. At the workplace, fatigue will adversely affect productivity (through impaired concentration or the need for rest pauses), absenteeism and safety. Fatigue can have consequences after work as well, as it can be a factor in drivers falling asleep at the wheel.

Night work tends to lead to a change in eating patterns, and may cause digestive disorders (ulcers and intestinal problems). Night workers may increase their consumption of fast foods, at the expense of proper meals. Consumption of coffee and "stay awake" drugs may increase. Further, since the body never really adjusts to night work, shiftworkers may not feel hungry when they finish their work, and may lose a regular eating pattern. Some shiftworkers have reported a general loss of appetite.

Finally, studies have suggested that workers on rotating shifts have a higher percentage of nervous disorders than those on day shifts. It is possible that sleep problems appear to be the symptom or cause of a more basic metabolic disorder.

Comparison of different shift rotations

Previously it had been assumed that long shift rotations were best for health, as they allowed the worker to adjust to each shift. There is now doubt about this, particularly since workers tend to revert to "normal" sleeping patterns on weekends and breaks. New thinking suggests that rapid shift rotation, with a maximum of three nights on night shift, is a better option. The worker is able to cope with a few consecutive nights of work without too much disruption to bodily systems or social structures. It appears that weekly rotation is the worst of all systems.

Recent studies suggest that employees working a 12-hour shift system have very few of the problems associated with shiftwork. The employees are able to maintain relatively normal sleeping and eating patterns, and have a large amount of time for family and social life. Typically, 12-hour shiftworkers have to cope with two nights of work every eight days, which are followed by a large period of rest. However, where the work involves a high degree of application, 12-hour shifts may not be appropriate.

¶11-140 Social effects

Night work or rotating shiftwork also has the potential to lead to some adverse social effects. These effects depend strongly on individual cases, however, and should not be generalised. The following list provides an indication of what may occur:

- disruption of family life, such as meals and contact with spouse/partner or children. The ability of the family to adapt to shiftwork is an important factor in determining whether the worker will be able to cope with the work
- participation in sport, clubs and groups is harder and, as the circle of friends is likely to be reduced, more individual forms of leisure may need to be pursued
- lessening of social facilities (theatre, film, television, licensing hours), all of which operate for the convenience of those who work nine to five
- employee relations problems may include inadequate communication of instructions from one shift to another and difficulty in obtaining general information on company activities, and
- the effects for women may be greater, due to their apparently greater deactivation rate than men, transport difficulties at night and the type of work they are frequently employed in, such as regulated assembly work. These issues do not imply that women are unsuited to night work, merely that employers should be aware of the different effects it may have on some individuals.

¶11-150 Stand-by periods

As mentioned previously, in some occupations a large amount of time may be spent "on call". Often this is in addition to the worker's regular load, with some financial compensation for the inconvenience of being on stand-by.

Stand-by periods mean that an employee is not able to rest and fully recuperate from work. Such periods may add to the employee's stress load and create anxiety. Workers on stand-by have reported poor quality sleep, most likely due to the possibility of interruption. Where sleep is actually interrupted by a call, it is unlikely that the worker will recover the lost sleeping time.

Being on call may cause the worker's normal life (interaction with friends and family) to be disrupted or to take a lower priority than work, leading to possible isolation and a further source of stress. Where the worker is called away frequently, the social problems discussed above in relation to shiftwork will result.

To properly accommodate stand-by periods into a working schedule, it is necessary to consider them as a workload. They are duty periods when the worker is not able to fully rest. Other periods should be allowed in order to ensure that the worker receives the necessary rest and recuperation to remain healthy in mind and body.

¶11-160 Reducing the ill effects

The following list provides guidelines as to how night work and shiftwork may be arranged in order to reduce the effects of some of the problems they cause.

1. Regular shifts may be preferable to rotating shifts, due to less fatigue and greater regularity of daily life. Where a rotating shift system is necessary, rapid rotation of morning, afternoon and night shifts with a maximum of three night shifts has produced significant improvements in health (demonstrated by both personal report and objective measurement) over workers engaging in the traditional three-weekly shift rotation system (see ¶11-130 for further details).
2. Workers should be given advice on how to cope with difficulties such as sleeping and meals (for example, how best to allocate "family time" and "sleeping time").
3. Shorter hours could be considered, either shorter shifts or preferably a shorter working week. This would better compensate for fatigue. Flexible working hours and/or additional annual leave could also be considered (note that a week's additional annual leave is a common award provision for shiftworkers).
4. Efforts should be made to reduce the number of people on night work to the essential minimum, through technology changes or attention to the distribution and operation of work tasks.
5. Retirement age could be lowered where superannuation schemes are provided. Susceptibility to fatigue increases with age, as does the period required to adapt to night work.
6. Improved facilities, such as transport, meal/cafeteria facilities and, where appropriate, beds.
7. Some people are attracted to night work initially because of the higher pay it offers, without considering whether they are really suited to it. Prior counselling may be a means of helping employees to adapt to the changes required by night work.
8. Employers should ensure that shiftworkers have access to employee services and facilities. For example, an answering machine could be provided so that workers can relay enquiries to the pay office/human resources department at night.

Tips on sleeping and eating for night workers

1. Plan sleeping times and do not change them. Make sure family and friends know these times.
2. If you have trouble sleeping enough in one stretch, try two: a longer sleep at the end of the shift, and a nap prior to returning to work.
3. Adjust your bedroom. Use heavy curtains, fan or air-conditioning, and no phone.
4. Do not eat a large meal just before sleeping. Also avoid alcohol, coffee and fatty foods that may prevent you from sleeping.
5. Do not substitute good meals with snacks and "junk foods".

Further references

Rogers, N, Roberts, T and Dawson, D, "Improving Shiftwork Management I: Understanding Shiftwork", *J Occup Health Safety — Aust NZ* 1997, 13(5), 429-437.

Reid, K, Roberts, T and Dawson, D, "Improving Shiftwork Management II: Shiftwork and Health", *J Occup Health Safety — Aust NZ* 1997, 13(5), 439-450.

Baker, A, Roberts, T and Dawson, D, "Improving Shiftwork Management III: Changing Hours of Work", *J Occup Health Safety — Aust NZ* 1997, 13(5), 451-460.

Shift Work, National Occupational Health and Safety Commission, 1991.

Guidelines for Workers and Management, Chamber of Mines and Energy of Western Australia, Inc, 1995.

Health and Safety Guidelines for Shift Work and Extended Working Hours, Australian Council of Trade Unions, 2002.

Shiftwork — How to Devise an Effective Roster, WorkCover NSW.

How to Manage Shiftwork, WorkCover NSW.

See also the websites of government OHS authorities and other OHS information providers such as those listed in the Appendices.

WORKPLACE VIOLENCE AND BULLYING

¶11-220 What is violence?

Violence is the unjust or unwarranted use of force and power. It can include verbal abuse, threats, harassment, physical assault, serious bodily injury and death. In its many forms, violence can occur in any occupation.

Workplace violence has been defined in various ways. Typical definitions include: "physical interpersonal aggression with the intent to harm or destroy" and "exertion of physical force so as to injure or abuse". Others define violence to include feelings of being threatened, abuse of managerial power, sexual harassment and gender issues. Common factors are that the violence occurs in conjunction with work, and that there is malicious intent and possibly physical injury. Other factors that could be included are psychological harm, and longer term outcomes of harassment and victimisation. The broader the definition, the more incidents will be captured.

Bullying

A distinction has been drawn between workplace bullying and workplace violence. Bullying is said to relate to internal conflicts between employees, including managers and supervisors, and is located within an organisation. Workplace bullying manifests as persistent and repeated aggressive behaviours that give rise to feelings of victimisation in the person who is subjected to it.

In contrast, workplace violence is seen in a broad range of behaviours and circumstances that can be present both inside and outside the workplace. Unlike bullying, violence can be a single event. It can also be ongoing, random, persistent or premeditated behaviour. Violence can be from an internal or external source and an assailant or perpetrator may be known or unknown to the victim(s). Workplace violence has been used to describe abuse, threats, rape, assault or murder of employees by members of the public, disgruntled ex-employees or service users.

Those responsible for violence or bullying may be individuals, groups of individuals or the organisation itself may have an aggressive culture. Organisational assault could include placing employees in dangerous work situations or exposing them to the risk of emotional trauma as a result of unfair practices relating to retrenchment, downsizing or redeployment. It could also include allowing a climate of bullying or harassment to thrive in the workplace.

If email and the internet are used to send confronting images or abusive messages to individual employees or groups of employees, some of these images and messages may also be considered as sexual harassment or so-called "cyber-bullying".

¶11-230 Incidence of violence

In this age of televised international belligerence, road-rage and computer games that foster aggression, it is perhaps not surprising that violence across the broader society is reflected in a corresponding level of violent incidents at work. The incidence of violence in the workplace has been investigated by a number of studies and surveys in recent years.

A United States study of security threats and management issues facing corporate America commented that "Workplace violence can be a most destructive and costly security threat in terms of its business impact. Work stops, reputations suffer. Productivity and morale decline. Expensive and distracting law suits crop up."

In Australia, surveys have found almost one in two Australians have been physically or verbally abused by someone with whom they work. In some occupations such as nursing, the level of violence is much higher still, and research shows that work-related violence can be very costly in terms of workers compensation claims. Acts of violence result in significant amounts of lost time, and very high costs for the most serious incidents.

Some possible reasons for this level of violence in a variety of businesses and service industries could include:

- criminals and drug-affected people robbing alternative targets because cash-handling organisations, such as banks and building societies, have increased their security arrangements
- people who are no longer eligible for government benefits and allowances as the welfare system is rationalised becoming increasingly desperate

- domestic disputes spilling over to the workplace
- psychiatric patients and juvenile offenders, being de-institutionalised and not provided with adequate support
- disgruntled employees (past and present) of downsized organisations seeking revenge, and
- a trend in society to accept the use of violence as a means to an end.

¶11-240 Likely targets

A wide variety of people may be the targets of work-based violence either directly or indirectly. These include:

- people in public contact situations (nurses, counter enquiry staff, sales personnel, hospitality staff and flight attendants)
- people working alone in community settings (real estate agents, security guards, taxi and bus drivers)
- staff handling cash or drugs (convenience stores, takeaway food outlets, video shops, petrol stations and pharmacies)
- those working within the legal system (judges, lawyers, child sexual assault workers, probation officers and police)
- private security staff (security guards and event management staff)
- those responsible for disciplining, relocating or terminating staff (HR personnel, security guards, supervisors and managers), and
- staff who are employed by organisations undergoing rapid and dramatic change (supervisors and managers).

A number of court cases reported in the media indicate a high incidence of violence inflicted against new members of organisations, such as apprentices and armed forces personnel, through dehumanising initiation and bastardisation practices.

¶11-250 Risk management approach

An employer will need to exercise a duty of care in the context of workplace violence. There are four different circumstances that apply to workplace violence:

- protecting workers against harassment, bullying, aggression and violence from fellow employees, past and present
- protecting workers from strangers, customers or estranged partners or family members
- protecting the employer's customers/clients from violent acts by its employees, and
- protecting customers/clients from other people coming onto the premises.

To fulfil this duty of care, workplace policies addressing the issue of violence must address the three interlocking issues of violence prevention, management and post-trauma support. Policies and guidelines must include education and training, forward planning and a quick response capacity for crisis situations, together with ongoing support for staff who experience violence in their workplace.

A risk management approach comprising the following elements is recommended.

Risk identification

This should be done with the cooperation of staff. A violence safety audit should be used to examine the following areas:

Policies and procedures — current codes of practice and safety policy; methods of induction; staff selection, training and supervision; monitoring of staff movements and current safety; staff support processes and availability; and incident recording and monitoring.

Location — working in isolated sites and working alone; disadvantaged neighbourhoods; difficult access to clients; reception areas; staff parking areas; interview rooms; and sites that may contain potential weapons.

Tasks — identification of staff pressure points; unacceptable task demands and rights and opportunities to abandon a task with unacceptably high risks.

Customers/clients — available information about a client's previous history; matching experienced staff with aggressive clients; clients made aware of their rights and complaints procedures; and inappropriate client referrals.

Staff — appropriate aggression management training; negative staff attitudes to clients or other staff; backup resources available to staff; alarm systems; opportunities to withdraw from high-risk tasks and locations; and identification of staff targeted by stalkers or former partners.

After-hours arrangements — arrangements for after-hours backup; and level of assistance provided for getting to and from work safely.

Risk assessment

Once the areas of risk are identified, risk assessment must take place. This determines the factors that are contributing to the risk of violence, the risk severity in each situation, the urgency of preventive action and the priority for action that must be assigned to each risk.

Once the risk assessment is completed, management should develop or modify its program for dealing with aggressive and violent situations. This program should include six interrelated aspects:

1. the organisation's operation
2. training issues
3. risk identification and assessment
4. incident management
5. post-incident management, and
6. ongoing monitoring and evaluation.

Risk control

Once the potentially violent situations have been identified and assessed, the next stage is to eliminate or control the risks by applying the control hierarchy of elimination, substitution, engineering controls, administrative controls and the use of personal protective equipment. Some suggestions follow.

Elimination — non-acceptance of dangerous clients or those affected by drugs or alcohol; close down the service or establish it elsewhere; outsource the service; screen out potentially violent employees; and implement a zero tolerance for violence policy.

Substitution — communicate by phone instead of direct face-to-face contact; select non-aggressive staff; use more experienced staff for difficult situations; and service non-aggressive clients.

Engineering controls — limit access; provide safe office/interviewing rooms; use electronic security devices; provide adequate external lighting; and provide safe and secure parking.

Administrative controls — develop clear policies and procedures; provide induction and ongoing training in aggression management; establish incident reporting and grievance management procedures; develop a crisis response plan; provide employee assistance programs; ensure staffing levels are adequate and rosters are realistic; and provide information about employers'/employees' rights and responsibilities.

Personal protective equipment — provide mobile phones and personal alarms where necessary.

By implementing the control strategies in cooperation with employees, violence in the workplace will be effectively managed, ensuring a safer and more secure environment for staff, clients and visitors.

Further information

Bowie, V, *Coping with Violence — a Guide for the Human Services*, revised 2nd edn, Whiting and Birch, London, 1996.

Occupational Violence — Guidelines for Queensland Health Employees, Workplace Health and Safety, Queensland, 2002.

Conference Crime Against Business, Melbourne, June 1998.

Violence at Work, International Labour Office, Geneva, 1998.

Being Bossed Around is Bad for Your Health, Australian Council of Trade Unions OHS Unit, April 2005.

Managing Violent and Potentially Violent Situations: a Guide for Workers and Organisations, Centre for Social Health, Melbourne, Victoria, 1997.

Grainger, C, *Violence: a Risk Management Handbook for Dealing with Violence at Work*, Miintinta Press, Brisbane, Queensland, 1994.

Violence at Work. A Workplace Health and Safety Guide, Queensland Department of Training and Industrial Relations, Division of Workplace Health and Safety, 1995.

Minimum Standards for the Prevention and Management of Occupational Assault, Victorian Department of Human Services WorkCover Unit, 1996.

Safety and Health Solutions. Violence in the Workplace, WorkSafe Western Australia, 1996.

See also the websites of government OHS authorities and other OHS information providers such as those listed in the Appendices.

AGE AND EXPERIENCE OF WORKERS

¶11-320 Issues relating to youth/inexperience or age of workers

Health and safety issues linked with the age or experience of individual workers need to be recognised and addressed. The risks to the young and inexperienced, as well as the potential problems with retaining older workers have been the focus of considerable attention in recent years.

One of the central principles of the modern approach to occupational health and safety is that the job should be made safe for the typical range of workers, rather than restricting personnel to only a select number who can cope with the working conditions. Within this context employers need to take into account the various characteristics of workers, including their age and experience, to ensure a safe and healthy working environment and systems of work that are safe for all workers.

Workplace risks to younger workers are apparent from the relevant national statistics, which indicate that more than 30 young people (under the age of 25) are killed at work each year in Australia. Thousands are injured, and a proportion of these are left with a permanent disability of some kind as a result.

The rate of injury to young men is particularly high — approximately three times higher than for young women. Data from the NOHSC indicate that young male workers, particularly apprentices, have a one in two chance of being injured at work each year, compared to a one in 18 chance of injury for older workers.

Accident rates for young people driving vehicles are known to be higher than rates for older people — hence the higher cost of car insurance if the vehicle is to be driven by a person under 25. This higher risk is reflected in the work context as well. Young men in particular can be over-confident in their ability to, for example, drive a forklift truck even though they may not have any experience with it and have not been trained in its safe operation.

Young workers are often unfamiliar with the workplace and its hazards, and they can be strongly influenced by peer pressure. They may feel, for example, a lack of interest or respect for safety considerations, and take unnecessary risks.

Youth and inexperience in the workplace can also make young workers more prone to other workplace dangers, such as bullying, victimisation or other oppressive, distressing or dangerous practices such as "initiation rituals". Employers must take factors such as these into account when defining safe systems of work. Training and supervision are particularly important in the case of young and inexperienced workers.

With older workers the issues are different. While governments try to encourage older workers to stay in employment, some employers are concerned that older workers may be less responsive, more difficult to train, have higher accident rates and take more time off work due to age-related illnesses. These invalid assumptions have significantly affected the job opportunities for older Australians at a time when there is a significant shortage of skilled workers.

Managing older workers, at least from an occupational health and safety viewpoint, should be no different than managing any other group of workers. In fact, as measured by many outcomes, older workers can be a very cost-effective resource for employers because of:

- accumulated skills
- lower training costs
- higher retention levels
- good work ethics
- excellent customer service skills
- low levels of sickness or absence due to personal or family member ill-health
- low likelihood of workers compensation claims
- significantly reduced absenteeism due to marital upsets, and
- lower absenteeism due to sport accidents and motor vehicle accidents.

¶11-330 Employer concerns

The most obvious aspect of young people in terms of their risk of workplace accidents is their physical immaturity. Developing bodies are more vulnerable in some ways to damage from manual handling of heavy loads, fatigue or other stresses.

Other characteristics of youth can also contribute to increased risk — consider the following.

- Unless taught otherwise, they tend to assume the risks are part of the job.
- Some working environments can be alien and intimidating to those who are not familiar with them.
- They may not recognise the potential for hazards in the working environment, due to their limited experience with it.
- They may be inhibited about asking questions, saying they do not understand instructions, or objecting if they think something looks dangerous.

- They may fear looking ridiculous or weak in front of their peers, and may resist wearing personal protective equipment (PPE) — especially if older co-workers neglect to wear PPE when appropriate.
- They have a greater tendency to "skylark" or take unnecessary risks.
- They may be experiencing a lifestyle change, because they have entered the workforce. This could mean more exposure to drug and alcohol use and more ready cash.
- They want to fit in, and do not always recognise their own limitations.
- They tend not to have proper sleep patterns and forgo nourishing meals due to social pressures.

Not all workers under 25 fit into the same mould, but employers need to consider the risks to which young people are exposed in the light of the above-listed tendencies. Another issue to take into account is the frequent tendency for young people to work while also undertaking a course of study — going straight from school or college to long shifts at work increases their risk of fatigue.

Because young people are often new to the workplace, employed on a casual or short-term basis and not familiar with all of the rules and procedures, they can be unaware of safety issues and of the employer's attempts to address health and safety. For example, they can be under-represented on workplace health and safety committees, and not oriented to the need to report risks. This again underscores the need for sufficient training and supervision of young people in the workplace.

In relation to older workers, employers may be concerned about physiological changes. Ageing by itself is not a disease, but some of the inescapable age-related physiological changes are associated with an increased prevalence of pathology. Age-related declines include:

- reduction in hearing acuity
- reduced visual acuity (even after adjustment with eyeglasses) as well as differing sensitivities to colours
- reduced maximal aerobic capacity and an increase in the rest time needed after performing physically demanding work
- reduced muscle strength, including hand grip
- reduced speed of processing new information
- decreased tolerance to heat or cold in the working environment
- increased prevalence of degenerative musculoskeletal changes, including arthritis and loss of bone mass
- increased prevalence of cardiovascular problems, both due to underlying disease such as hypertension, as well as the possible effects of medications
- increased risks of undetected metabolic diseases such as diabetes
- increased prevalence of dementia, and

- increased difficulty in coping with shiftwork and extended working hours.

Among this array of potential age-related disorders, a large degree of individual variation will occur. The healthy worker effect is especially prominent in older workers, with clear evidence of self-selection. Fitter, older workers often choose to work to an older age. Those with serious underlying health problems do not. A physically active, healthy 70-year-old man may have much greater physical strength than a 45-year-old woman.

These individual differences increase with age, and clearly require fitting the individual to the job. This is essential for workers aged over 50. In addition to assessing the physical, mental and social requirements of the task, a responsible employer will focus on the individual's work capacity or functional age, rather than just rely on the very crude indicator, chronological age.

Surveys undertaken in Europe, the US and Australia have consistently shown that employers recognise the value of older workers, and acknowledge their lower absenteeism, greater work ethic, thoroughness and reliability, etc, but consistently (and erroneously) believe that older workers have more accidents and cannot be retrained for new tasks.

Though this is not always true, accident rates among older workers are in fact higher in certain classes of employment. Older workers have higher accident rates due to slips and falls and are involved in a higher percentage of transportation-related job fatalities. However, in the vast majority of post-industrial Australian workplaces, these potential problems can be avoided.

With proper floor design, non-slip surfaces, adequate guard rails and appropriate use of safety harnesses, the working environment can be made safer for all workers, not just those younger or older than some arbitrary figure. The transportation fatality rates, especially in Europe and the US, are largely caused by the much higher percentage of older owner-drivers driving their own semi-trailers and operating as independent contractors. Lifestyle issues contribute to a higher accident rate in this group of workers. These issues include high levels of cigarette smoking and hypertension, relatively low driving skill levels, older mechanical equipment, excessive tiredness and the financial necessity to keep working to demanding schedules to earn a decent return.

Health promotion and screening programs, coupled to rigorous enforcement of safe working hours, use of log books and the recent introduction of satellite tracking of heavy transport vehicles, should all lead to lower accident rates in this group of workers.

Concerns about the training of older workers have a long history. However, it has been demonstrated that older people can learn as fast as younger ones if appropriate teaching methods are adopted.

A US report examined the cost/benefit of retraining older workers versus replacing them with younger workers. The analysis showed that the cost of retraining 20 older engineers (half of them in their 50s) was only one-third of the cost of replacing them with younger employees. Another study concluded that forcibly retiring workers aged over 55 was over 30% more expensive than retraining and continuing to employ the older workers.

The US National Foundation for Occupational Health Research has concluded that there is no clear-cut evidence that older workers show a significant decline in their skills or capacities, and in certain areas they can match or outperform their younger counterparts.

¶11-340 Managing risks for young/inexperienced and older workers

The provision of adequate training and supervision is the foundation of managing the health and safety of young or inexperienced workers.

Induction training as soon as a young person joins an organisation — before they commence duties — should build on any vocational training the employee has had at school or TAFE and should introduce the worker to the health and safety risks inherent in the job.

It is important to focus on the job the worker will be performing, rather than just the employer's premises. For example, young bicycle couriers have been hit by cars, or young people injured in fast-food home delivery. Though the injuries did not happen on the employers' premises, risks on the road are still part of the full set of risks to which the worker is exposed when carrying out the employer's business, and so should receive just as much attention during induction training.

All workers, and especially young workers, should receive induction training that covers the following matters:

- the organisation's health and safety policy
- hazards present in the workplace, or involved in work procedures
- relevant safe work practices
- behaviour likely to lead to danger or injury
- any documentation the worker should be aware of (such as operating manuals and material safety data sheets)
- maintenance and service routines (for example, when to request maintenance and who to ask)
- any special safety information needed (such as precautions for working under certain conditions, how to use emergency stop buttons or other safety devices, and rules about use of drugs and alcohol)
- the types of personal protective equipment needed, together with information on fitting, using, maintaining and storing this equipment
- emergency procedures
- reporting requirements for injuries or incidents, and
- where relevant, details on how accidents have occurred in the past involving the same work process.

On-the-job instruction in the particular tasks to be performed is an essential follow-up to formal or general training. This should recognise that young persons cannot be expected to be aware of all the dangers that more experienced or mature workers might know and avoid. The question of competence to drive or operate machinery is especially important. Young men in particular need to understand that they are not free to use a vehicle if they have not been trained in its safe use — even if the key has been unwisely left in the ignition.

Supervision of all employees, especially the young, in the first few weeks of work, is vital to ensure that training and instruction have been understood and safe working practices are being adhered to. Young workers may adopt role models among older or more experienced workers. The example presented by other workers and peer pressure to emulate other workers' practices make it clear that the full context of training has to be considered. Formal training can easily be undermined if it is contradicted by the "informal training" that happens when the young worker observes how the other workers do things.

The importance of adequate instruction and supervision for young workers highlights the need for adequate supervisor training. Supervisors must be competent not only in the quality of initial instructions to workers, but also in terms of their follow-up observations of how young workers perform the tasks. They also need to recognise the effects of peer pressure and general work practices. Supervisors must be able to achieve the same level of compliance with safe work practices among older workers as they expect from younger workers.

Managers in turn need to assess the adequacy and effectiveness of supervisor training, as well as the suitability of induction and on-the-job training. They also need to monitor compliance with safe operating procedures in which workers have been trained.

Managing older workers should not be unduly difficult — being a good people manager implies an ability to ensure that an organisation's needs are effectively married to its workers' needs to ensure optimum outcomes for all parties. In practice, however, managers need to be aware of some of the sensible and cost-effective workplace adjustments that should be made to accommodate older employees.

Vision screening

At the simplest level, employers with visually demanding tasks should consider the implementation of visual acuity screening tests for employment applicants. Since eyesight gradually deteriorates with age, a significant percentage of older workers will need to have their vision corrected to perform their jobs safely and efficiently.

While it may be obvious to everyone that an older driver of a heavy transport vehicle, eg a petrol tanker, will need excellent (corrected) eyesight, many other employees in the jobs working with computers and other visual displays might also need sight correction. Frequently, vision testing only focuses on reading distance and infinity. However, workers need to be able to focus clearly on computer screens or other displays at the viewing distance which may be 500 mm to 800 mm from the eye. Traditional refractions will not compensate for this viewing distance, and employees with prescription eyeglasses need to advise their optometrist or ophthalmologist of the focal distances that they will be working at.

Employers of large workforces in some industries in the US and Europe provide free eyesight screening, and either free or greatly subsidised prescription eyeglasses to help their employees see most clearly at the required viewing distance. These in-house programs are of benefit to all workers, but especially older workers whose uncorrected eyesight accommodation will have deteriorated with age. This also assists the employer in reducing the number of errors resulting from poor recognition of work.

Health promotion programs

Health promotion programs can substantially benefit all employees, and especially older workers. Healthy older workers will greatly benefit from periodic screening programs covering areas such as glaucoma detection, diabetes screening, hypertension screening and screening for certain types of cancers, especially those in the breast, prostate and colon. These programs have widespread community acceptance and may also significantly benefit workplace morale and the corporate image. The programs have been shown internationally to be cost effective. They can reduce absenteeism by assisting workers to attend to their health.

If a company has a long-term and ongoing investment in health promotion at the workplace, a substantial amount of valuable health data can be collected. This enables early detection of departures from normal with consequent improvements in treatment and health outcomes. Ultimately, this will improve worker productivity through lower absenteeism, and greater feelings of wellbeing.

One controversial area for employers to consider is the issue of influenza vaccination for workers. National Health and Medical Research Council data basically suggest that even healthy older workers will benefit from yearly influenza vaccination. The problems are not so much just being sick with influenza for a few days, but the serious health consequences such as pneumonia that can result from a bout of influenza. While this is of major concern to older workers, many younger workers are also at risk, especially if they have pre-existing chronic diseases or some immune system compromise. This is a good example of where designing a program for older workers can have significant benefits for all workers.

Shiftwork

Many workers liken the effects of shiftwork to chronic jet lag. Numerous studies have shown that older workers do not tolerate forced shiftwork as well as younger workers. There are many reasons for this, including difficulty in sleeping, adjusting to altered day/night cycles, gastrointestinal upsets and circadian dysrhythmia affecting conditions such as diabetes mellitus and hypertension, etc.

Many older workers select themselves out of consideration for shiftwork jobs. Their financial needs are usually lower, and the extra money from shift loadings, etc, is just not worth the side effects of shiftwork. It is suggested that, wherever possible, shiftwork requirements be met entirely with volunteers and that older workers be given reasonable consideration in relation to requests for transfer to permanent day work. Shift rotation and length should be structured to reduce the impact on individuals.

Training and retraining

Old dogs can learn new tricks, and the principles of adult learning have been understood for nearly a century. While our technology is changing rapidly, older workers in particular may feel marginalised when their 20-plus years of experience suddenly become valueless. The occupation of typesetting is an example. This was a trade that had been fairly stable for 500 years, yet in less than one generation, the technology requirements of today's publishing have meant that the manual setting of type has become a lost art.

What is needed in retraining workers, particularly older workers, is the opportunity to experiment, to make mistakes, and learn by doing. In the case of computer technology, personal computers have been around for only a little over 20 years, and the world wide web for even less. Mature professionals may therefore not have been exposed to these technologies at home, school, TAFE or university.

Quality training at an appropriate pace, frequent reinforcement and support, and the ability to experiment at home or at work in a low-risk environment will help older workers to become proficient in the use of the newer technologies. Once acquired, these new skills will enable older workers to successfully meet the technology challenges of today's business environment.

Further information

Gardner, I, "Older Workers and Safety", in *OHS Alert* Issue 5, CCH Australia, June 2005.

Age Counts: an Inquiry into Issues Specific to Mature-age Workers, House of Representatives Standing Committee on Employment, Education and Workplace Relations, Canberra, 2000.

Patrickson, M and Hartmann, L (eds), *Managing an Ageing Workforce*, Australia, BPP Publishing, 1998.

Are Young Workers a Greater Risk? Factsheet no 2, Workplace Health and Safety Division, Queensland Chamber of Commerce and Industry, 1999.

Young Workers, Work Experience and Work Placement Students — Don't Let Their First Day Be Their Worst Day, WorkCover NSW, 1997.

See also the websites of government OHS authorities and other OHS information providers such as those listed in the Appendices.

APPENDIX A

RESOURCES: ORGANISATIONS

Page

Introduction

Throughout this book, it has been emphasised that occupational health and safety covers many areas and involves a large number of technical issues. Multidisciplinary knowledge is required to cover all aspects of the subject. The aim of this book is not to cover complex issues in full detail, but to act as a starting point for practitioners to seek further details for themselves.

Every organisation will face different problems and circumstances at its own workplace and will therefore have differing needs for information in order to resolve problems. Different solutions may also be appropriate.

In recognition of all these factors, the purpose of these appendices is to provide a guide to the sources of both general and specialised assistance available to organisations. This Appendix A describes the range of government and private organisations in the health and safety field; Appendix B lists some of the available information and training resources; and Appendix C describes the roles and qualifications of a variety of OHS practitioners.

Government organisations

Commonwealth

Australian Safety and Compensation Council

The Australian Safety and Compensation Council (ASCC) has been established to promote a national approach to workplace safety and workers compensation in Australia. Its main role will be to coordinate research and provide policy advice to the Workplace Relations Ministerial Council — which comprises the Federal Workplace Relations Minister and his or her state and territory counterparts. The ASCC is intended to be a key forum for the promotion of greater national coordination, consistency and uniformity amongst the various jurisdictions.

The ASCC replaces the National Occupational Health and Safety Commission. Visit the website at www.ascc.gov.au.

National Occupational Health and Safety Commission (NOHSC) (formerly known as Worksafe Australia)

The NOHSC was established as a statutory body in 1985. Based in Canberra, it was a tripartite organisation with input from government, employers and unions.

Under the *National Occupational Health and Safety Commission Act 1985*, it was charged with a range of specific functions. These included: formulation of policies and strategies; encouragement of cooperation between Commonwealth, state and territory governments and groups and individuals; declaration of national standards and codes of practice; collection, interpretation and dissemination of information; conduct of inquiries; provision and assistance with training in occupational health and safety matters; publication of reports and papers; provision of grants; support of research and testing; and encouragement of the application or use of the results.

The NOHSC developed national standards and codes of practice that have considerable weight as guides to acceptable practice. Though its role has been taken over by the Australian Safety and Compensation Council, the standards, model regulations and guidance material it produced represent the core of established good practice in OHS in Australia.

The National Occupational Health and Safety Commission publications (see Appendix B) are available through NOHSC's website at www.nohsc.gov.au or ASCC's website at www.ascc.gov.au.

Comcare Australia

Comcare Australia is the workers compensation insurer for the federal government, providing health and safety, rehabilitation and compensation services to federal employees. It has its head office in Canberra and offices in the other states and territories. It administers the *Occupational Health and Safety (Federal Employment) Act 1991*, and is answerable to the Federal Minister for Workplace Relations and Small Business.

Comcare does not have its own inspectorate, but manages its inspection functions through cooperative arrangements with each of the states and territories. Visit their website at www.comcare.gov.au.

National Industrial Chemicals Notification and Assessment Scheme

The National Industrial Chemicals Notification and Assessment Scheme (NICNAS) is a statutory scheme located within the Office of Chemical Safety, in the Australian Department of Health and Ageing. It assesses industrial chemicals for their health and environmental effects prior to their use in Australia, and makes recommendations for safe use. The NICNAS website is at www.nicnas.gov.au.

The Australian Radiation Protection and Nuclear Safety Agency

The Australian Radiation Protection and Nuclear Safety Agency (ARPANSA) is a federal government agency charged with responsibility for protecting the health and safety of people, and the environment, from the harmful effects of ionising and non-ionising radiation. It was established in 1998 from a merger of the Australian Radiation Laboratory (ARL) with the Nuclear Safety Bureau (NSB).

Specific activities undertaken at ARPANSA include standards calibration, regulation of nuclear facilities and ionising radiation sources, a personal radiation monitoring service, and monitoring of environmental radiation. Details of these and many other activities can be found on their website at www.arpansa.gov.au.

National Acoustic Laboratories

The National Acoustic Laboratories (NAL) undertakes scientific investigations into hearing, hearing habilitation and rehabilitation, and the prevention of hearing loss. It is the research division of Australian Hearing, a Commonwealth government authority under the Department of Human Services. Visit NAL's website at www.nal.gov.au.

CRS Australia

Formerly known as the Commonwealth Rehabilitation Service, CRS Australia delivers vocational rehabilitation services to Australian citizens or residents who have an injury, disability or health condition. As part of the Department of Human Services, it assists people who have a disability or injury to get a job or return to their job. This service is free to people receiving income support payments from Centrelink. It also provides occupational health and safety consultancy services to employers. Visit the website at www.crsrehab.gov.au.

National Health and Medical Research Council

The National Health and Medical Research Council (NHMRC) is a fully independent statutory agency within the Health and Ageing portfolio. The role of the NHMRC is to support excellence in research, health advice, and research and health ethics that leads to improvement in the health of all Australians. It is Australia's leading expert body promoting the development and maintenance of public and individual health standards. Visit the website at www7.health.gov.au/nhmrc/index.htm.

Australian Nuclear Science and Technology Organisation (ANSTO)

ANSTO is Australia's national nuclear technology and research organisation. The organisation is located at Lucas Heights, on the outskirts of Sydney, where it operates two research reactors. The organisation is heavily involved in occupational health and safety, providing expertise in all areas of nuclear safety and conducting a wide range of general safety courses. ANSTO can be found at www.ansto.gov.au.

State government departments

Most of the relevant legislation is state government-sourced. The principal authority responsible in each state and territory is given below:

NSW

WorkCover Authority of NSW

Information centre: 13 10 50

Website: www.workcover.nsw.gov.au

Vic

Advisory service: 1800 136 089

WorkSafe Victoria (the health and safety arm of the Victorian WorkCover Authority)

Website: www.workcover.vic.gov.au

Qld

Workplace Health and Safety, Department of Industrial Relations

Advisory service: 1300 369 915

Website: www.whs.qld.gov.au

SA

WorkCover Corporation of South Australia

Advisory service: 13 18 55

Website: www.workcover.com

Note that WorkCover's OHS functions will be taken over by SafeWork SA from 1 January 2006.

WA

WorkSafe Western Australia

Phone: (08) 9327 8777

Website: www.safetyline.wa.gov.au

Tas

Workplace Standards Tasmania

Phone: 1300 366 322

Website: www.wst.tas.gov.au

ACT

ACT WorkCover

Phone: (02) 6205 0200

Website: www.workcover.act.gov.au

NT

NT WorkSafe

Phone: 1800 019 115

Website: www.worksafe.nt.gov.au

Other departments may administer legislation on subjects such as mining, water, noise and air pollution, disposal of toxic wastes, approval for electrical equipment, agriculture and poisons. Note that some matters may fall within the administrative control of local government councils.

Other organisations

The list of organisations here is not meant to be exhaustive. In addition to the ones listed here, there are specialist organisations and a wide range of consultants who advise on particular topics such as lighting, acoustics, stress management techniques, harmful substances, air-conditioning, etc. There are also a number of organisations that provide useful software.

Standards Association of Australia (Standards Australia)

The Standards Association of Australia is a non-government, independent body incorporated by Royal Charter. It is known by its trading name of Standards Australia, and is referred to as such throughout this book.

The association's main function of investigation, development and recommendation of standards includes standards relevant to occupational health and safety. These standards are prepared by a process of voluntary consensus and many have been called up in state legislation. When Australian standards are called up in legislation they become laws. In other circumstances, standards are regarded as authoritative or, at least, as excellent guidelines to the appropriate quality or procedure.

Australian standards are distributed by SAI Global Limited under licence. Copies of all Australian standards are available over the phone from the SAI Global's Customer Service Centre, or from the website, www.standards.com.au. Charges are based on the size of the particular standard.

The website includes a focus on codes and standards that affect occupational safety, and a magazine, *The Global Standard.*

National Safety Council of Australia (NSCA)

The National Safety Council of Australia (NSCA) is a single incorporated national not-for-profit organisation, funded by a combination of members' subscriptions and discounted fees for services provided to members and non-members.

The council has branch offices throughout Tasmania, New South Wales, Victoria and Queensland, with formally affiliated offices in South Australia, Western Australia and the Northern Territory. It offers professional consultancy services and guidance on the initiation and implementation of occupational health and safety management programs. It is the largest organisation of its type within Australasia, and its main aim is the continuous improvement of occupational health and safety in both the private and public sectors.

Services offered include health and safety surveys, audits, scientific services, quantified risk assessments, a wide range of management and employee training, risk management programs and behavioural change processes. The council also offers several awards as incentives to organisations and individuals. The winners of these awards are published in NSCA's national magazine *National Safety*. Visit the website at www.nsca.org.au.

Safety Institute of Australia (SIA)

The Safety Institute of Australia Inc is a voluntary organisation with a membership drawn from people carrying out occupational health and safety functions in a wide range of industrial and government employment. The institute aims to promote the science and practice of accident prevention, to facilitate the exchange of information and to encourage the development of safety as a profession. Besides a federal council that coordinates activities, there are divisions in each state and the ACT. Visit the institute's website at www.sia.org.au.

Australasian Faculty of Occupational Medicine (AFOM)

The faculty aims to establish and maintain standards of training and practice in occupational medicine in Australia. To this end, it runs intensive training programs enabling medical doctors to specialise as occupational physicians. Visit the website at http://afom.racp.edu.au/.

Human Factors and Ergonomics Society of Australia

The Human Factors and Ergonomics Society of Australia Inc (HFESA) is the professional organisation of ergonomists and human factors specialists in Australia. Its purpose is to promote the principles and practice of ergonomics throughout the community. It runs annual conferences and each state has a professional development program. HFESA is one of 36 federated societies worldwide that comprise the International Ergonomics Association. Visit the website at www.ergonomics.org.au.

Australian Institute of Occupational Hygienists (AIOH)

The Australian Institute of Occupational Hygienists aims to promote its profession, to improve the knowledge and practice of its members and to represent them at national and international levels. The institute maintains a register of professional occupational hygienists. Visit the website at www.aioh.org.au.

Australian and New Zealand Society of Occupational Medicine (ANZSOM)

The Australian and New Zealand Society of Occupational Medicine is a professional organisation which provides a focal point for the advancement of knowledge for those registered medical practitioners involved in occupational medicine. The society seeks to educate people in the proper practices of occupational health. Visit the website at www.anzsom.org.au.

Australian College of Occupational Health Nurses (ACOHN)

The Australian College of Occupational Health Nurses is the professional body for occupational health nurses in Australia. It was formed to meet the needs of nurse practitioners in industry, commerce and government. The purpose of the college is to represent and promote the professional standing, diverse function and competence of its members and encourage excellence in the practice of occupational health. This is achieved through education, policy advice and development, research, exchange of ideas and conferences and meetings. Visit the website at www.acohn.com.au.

Fire Protection Association Australia

Fire Protection Association Australia is a non-profit, technical and educational organisation with the aim of promoting continual improvement in fire safety. The association maintains a technical and educational reference library and a film library of instructional films that are loaned free to members. Training courses and seminars are conducted, and bulletins and a journal are produced. Visit the website at www.fpaa.com.au.

St John Ambulance Association and Australian Red Cross Society

Both organisations conduct certificate courses in first aid, that are acceptable as an approved level course under safety legislation. Courses can be arranged within an organisation (if numbers are sufficient) or for a combined group. Visit the websites at www.first-aid.com.au and www.redcross.org.au.

Australian Council of Trade Unions (ACTU)

The ACTU has adopted an official policy on occupational health and safety, that covers ACTU involvement in legislative change, research, education and training, managerial action and direct union activity. It has also established an Occupational Health Unit. The unit publishes a regular newsletter, as well as guidelines on particular subjects, such as VDUs and manual handling. Visit the website at www.actu.asn.au.

State labor councils and unions

State labor councils and a number of unions have OHS/welfare departments that provide information on OHS issues relating to their members. This information is often distributed through newsletters or publications. At times these bodies will also undertake or fund studies into particular OHS concerns of their members. Labor councils' and unions' websites are linked with the ACTU's website.

Trade Union Training Authority (TUTA)

TUTA is a statutory authority which plans, develops and conducts courses for trade union officials, delegates and members. Some of these courses deal with occupational health and safety either solely or in part. Visit TUTA's website at http://www.alphalink.com.au/~mojo/xtras/tuta/Hmpage1.htm.

Workers Health Centres

Workers Health Centres generally specialise in the provision of occupational health and safety services to union members. The centres are also involved in the promotion of safer workplaces. Centres are located in Sydney, Wollongong, Newcastle, Melbourne and Brisbane. Visit the Workers Health Centre in Sydney at its website www.workershealth.com.au.

Australian Chamber of Commerce & Industry (ACCI)

The Australian Chamber of Commerce & Industry is the largest single organisation representing industry and commerce at a federal level. The ACCI is the employer representative on the ASCC.

The ACCI Occupational Health and Safety Facility provides information on occupational health and safety to employer organisations with literature and database searches. They develop and distribute information on occupational health and safety issues, and distribute draft standards and codes of practice for information and comment by employers. Visit the website at www.acci.asn.au.

Employee Assistance Professional Association of Australia

For information from the Employee Assistance Professional Association of Australia, visit its website at www.eapaa.org.au.

Employee health and welfare assistance programs are available for workplaces from private organisations which provide this service. The services available include consulting, training and counselling on problems in the workplace, including drug and alcohol and work-related post-accident trauma. For further information, contact the occupational health and safety or industrial relations authority in your state or territory.

See also ¶9-650 for more information on EAPs.

APPENDIX B

INFORMATION AND TRAINING RESOURCES

References

It is not possible to provide an exhaustive list of available OHS information. The following is a guide to a range of useful general references, books, periodicals and websites.

Australian Occupational Health & Safety Law (3 vols), CCH Australia Limited, Sydney

This online reporting service (also available in loose-leaf hard copy) emphasises the legislative side of occupational health and safety. Legislation is reproduced either in full text or as extracts. It contains commentary on legal liability, planning, promotion and rehabilitation, administration and enforcement and particular health problems. It also includes a comprehensive reference and advisory services section.

Managing Occupational Health & Safety, CCH Australia Limited, Sydney

This online reporting service (also available as a loose-leaf) aims to provide a guide for conducting and administering occupational health and safety at the workplace. The manual does not set out legislation, but aims to provide a practical approach to dealing effectively with occupational health and safety. The manual's sections include a guide to the human body, hazard control, training, first aid, emergency planning and record keeping, as well as dealing with a variety of specific occupational safety problems (hearing conservation, electrical safety, etc).

National Occupational Health and Safety Commission publications

The National Occupational Health and Safety Commission (NOHSC) published a range of national standards and codes of practice dealing with specific health and safety issues. In addition, there are information brochures, research reports, resource kits and workplace guides.

At the time of writing, these publications are not available in hard copy, though print versions may again be available at some time in the future. They may be viewed in libraries with special OHS collections, or online via the federal government's publications website at www.publications.gov.au. As the commission's responsibilities have been taken over by the Australian Safety and Compensation Council, the information is also available through the ASCC's website at www.ascc.gov.au.

Encyclopaedia of Occupational Health and Safety, ILO, Geneva

The fourth edition of this comprehensive reference work was published in 1998. Written by 900 specialists, it covers a huge range of occupational health and safety topics. Copies are held by the Commonwealth and state government departments concerned, as well as by many voluntary organisations. For inquiries or availability, contact Hunter Publications, Melbourne.

Periodicals

A comprehensive list of international periodicals appears in the reference section of CCH's *Australian Occupational Health & Safety Law* (see above). The following is a list of useful Australian journals.

- *The Journal of Occupational Health and Safety — Australia and New Zealand*, CCH Australia Limited
- *National Safety*, National Safety Council of Australia (some state branches of the council also publish their own newsletters)
- *Occupational Health News*, Thomson CDP, Sydney
- *Ergonomics Australia*, Human Factors and Ergonomics Society of Australia, Qld
- *Fire Australia*, Fire Protection Association Australia
- *The Global Standard*, SAI Global Ltd, Sydney
- *WorkCover News*, WorkCover Authority of NSW, Sydney
- *Health and Safety Bulletin*, WorkCover Corporation of South Australia, Adelaide
- *Safetyline*, WorkSafe Western Australia.

Books

CCH's *Australian Occupational Health & Safety Law* (see above) contains a comprehensive book list, divided into subject titles. *The Journal of Occupational Health and Safety — Australia and New Zealand* (see above) features book reviews including summaries of government reports.

A selection of useful general titles is provided below.

Bluff, L, Gunningham, N, Johnstone, R, *OHS Regulation for a Changing World of Work*, The Federation Press, Sydney, 2004.

Ellis, N, *Work and Health: Management in Australia and New Zealand*, Oxford University Press, Melbourne, 2001.

This book has been described as a ground-breaking Australasian OHS textbook. Its scope is comprehensive and it is supported by many case studies.

Frick, K, Jensen, P, Quinlan, M and Wilthagen, T, *Systematic Occupational Health and Safety Management*, Pergamon, 2000.

This book is a critical discussion of the empirical evidence for the effectiveness of OHS management systems.

Bohle, P and Quinlan, M, *Managing Occupational Health and Safety: a Multidisciplinary Approach* (2nd edn), Macmillan Education Australia, 2000.

This informative and comprehensive textbook establishes OHS as an issue that needs to be considered in more ways than scientific and technical.

Evaluation of Human Work: a Practical Ergonomics Methodology (edited by Wilson and Corlett), Taylor and Francis, London, 1995.

An ergonomics text aimed specifically at applying ergonomics in the workplace. It comprises 34 chapters written by 38 contributors.

Websites

In addition to the websites of government OHS authorities and other organisations already listed in Appendix A, many other organisations provide health and safety information about specific issues and risk management strategies for OHS problems. The sites listed below will in many cases have links to other relevant sites.

For the text of legislation: www.austlii.edu.au.

The Canadian Centre for Occupational Health and Safety at www.ccohs.ca.

(US) Department of Health and Human Services, Centers for Disease Control & Prevention at www.cdc.gov/niosh.

International Labour Organisation, OHS Information Centre (CIS) at www.ilo.org/dyn/cisdoc.

World Health Organization, Occupational Health, at www.who.int/occupational_health.

(UK) Health and Safety Executive at www.hse.gov.uk.

Finnish Institute of Occupational Health at www.occuphealth.fi/Internet/English.

Internet Safety Resource (contains a listing of over 2000 health and safety internet-based resources) at www.christie.ab.ca/safelist.

MSDS Solutions (which offer over one million MSDSs free of charge, as well as links to other chemical information sites) at www.ilpi.com/msds.

The University of Queensland's Minerals Industry Safety & Health Centre at www.mishc.uq.edu.au.

Industrial Foundation for Accident Prevention (WA) at www.ifap.asn.au.

New Zealand Department of Labour's OHS Service at www.osh.dol.govt.nz.

Audio-visual aids

Videos covering occupational health and safety topics can provide very powerful communications. Interest in the area has seen a rapid expansion of releases produced independently and in association with various interested organisations.

The National Occupational Health and Safety Commission built up a comprehensive collection of OHS videos and publishes a catalogue of these, which can now be accessed via the library of the Australian Safety and Compensation Council at www.ascc.gov.au.

Many of the other organisations listed in this appendix will be able to provide videos on general or particular subjects, or else be able to advise where such items may be obtained. In addition, there are private organisations that rent and sell safety videos.

Training and tertiary courses

Training courses can be divided into three groups, government-run courses, those conducted by private organisations, and courses run at educational institutions. As with other areas of occupational health and safety, there has been a rapid expansion in training and courses for all levels of interest. The various government bodies and private organisations noted in the previous paragraphs of this appendix and Appendix A may be contacted for information. In addition, details of training and courses are featured from time to time in *The Journal of Occupational Health and Safety — Australia and New Zealand* (see "Periodicals", above).

Tertiary courses in occupational health and safety or risk management are available from many universities and TAFE colleges. The course information office at institutions in local areas can provide detailed information on course content and registration requirements. Many courses are part-time, lasting from two to four years.

Other CCH products in the area of OHS

OHS Solution Finder

OHS Solution Finder is a powerful tool combining the search capabilities of internet technology with CCH's information database of expert knowledge and extensive resources. Its aim is to make it easy to find the answers to particular OHS questions.

Every piece of legislation, case law, commentary and regulatory release in CCH products linked with this service has been studied and linked to relevant topic areas, allowing you to cross-reference the extensive CCH library at topic level. The end result is that you get precise answers you need — faster.

Online reporting services (also available in loose-leaf hard copy)

- *Australian Workers Compensation Guide*
- *The Hands On Guide — OHS Legal Guide*
- *The Hands On Guide — OHS Manager*
- *The Hands On Guide — School Health and Safety*
- *Victorian Accident Compensation Practice Guide*

Loose-leaf reporters

- *Plant Safety: Managing Plant Hazards in the Workplace*
- *Hazard Alert: Managing Workplace Hazardous Substances*
- *Hazard Alert: Hazardous Substances Induction Training Pack*
- *Laboratory Safety Manual*

- *The Hands On Guide — Risk Management*
- *The Hands On Guide — Managing Contractors*
- *The Hands On Guide — Manual Handling*
- *Safe Mining*

Training

- *Occupational Health and Safety Training Kit*

Books

- *Safety, Culture and Risk*
- *Australian Master OHS and Environment Guide 2003*
- *Guidebook to Workers Compensation in Australia*
- *Occupational Health and Safety Laws in Australia*
- *Understanding New South Wales Occupational Health and Safety Legislation*
- *Lessons from Longford — the Esso Gas Plant Explosion*

APPENDIX C

OHS PROFESSIONALS

When to engage an expert

Special problems and staff limitations may give rise to the need to engage an outside expert. Using consultants has the advantage of bringing in more specialised skills than the organisation generally has access to. Consultants are separate from the workplace and are able to provide a fresh approach and an objective assessment.

Remember that when the outside assistance needed is more in the nature of general advice rather than a detailed assessment, various government authorities may be able to help. The major organisations are listed in Appendix A.

Engaging a consultant

Preparation prior to engaging a consultant can save time and money. Remember that you are a consumer and the consultant is offering a service.

The following steps can be adopted:

1. Define the problem and what the consultant will be required to do.
2. List those consultants who may be able to assist.
3. Contact the consultants by phone and compile a short list of those who are suitable.
4. Call for written proposals from those on the short list.
5. Evaluate the proposals — cost is not the only criterion.
6. Draw up a contract specifying the exact nature of the project required — where the project is lengthy, specify interim reporting requirements and the ability to alter the project.

The occupational health nurse

An occupational health nurse is a qualified nurse who has undertaken specific studies in occupational health. The Australian College of Occupational Health Nurses has issued a set of guidelines for the employment of occupational health nurses in Australia. The following outline is based on material in that publication.

Duties

The duties of each occupational health nurse will vary according to the size of the organisation, the number of employees, the nature of the work and associated hazards, the location and, of course, the number of nurses and other staff employed in the health service. These duties are all directed towards the prevention of occupational illness and injury and the promotion of physical and mental health.

The list below provides an indication of the range of duties.

The trend in the role of occupational health nursing in recent years has been to move more towards preventive functions. Many nurses will be involved in designing and implementing workplace programs in addition to the more traditional "band-aid" role.

Duties of an occupational health nurse

1. Pre-placement health assessment of new employees or employees changing their job within the organisation.
2. Monitoring employees' health routinely, for statutory purposes or for those at risk (eg executives, employees exposed to noise).
3. Assessment of employees who return to work after illness or injury, or who request permission to leave the premises because of illness.
4. Surveillance of employees who need special attention, such as those with known medical conditions, pregnant women, young people, persons with disabilities and workers exposed to special risks, such as lead.
5. Provision of immediate and follow-up nursing management in cases of illness or injury occurring at work.
6. Referring employees where appropriate to their own general medical practitioner, hospital, specialist or the company's medical adviser.
7. Giving advice to individual employees on health and hygiene.
8. Organisation of first aid facilities and services.
9. Supervision and training of first aiders.
10. Establishment and maintenance of accurate, complete and confidential health records, that may include information for workers compensation requirements.
11. Preparation of appropriate statistical records of all attendances at the health centre or services provided.
12. Preparation of reports for management including on-site assessments and nursing activities.
13. Involvement in risk management programs, safety surveys, work environment visits, and safety and health education activities.
14. Liaison, cooperation and communication with other personnel, departments, trade unions, community organisations and programs concerned with employees' health, safety and welfare.
15. Health and safety research.
16. Involvement in counselling and rehabilitation.
17. Recommendation of other OHS specialists who may be appropriate.

Qualifications and training required

In addition to basic training and some practical experience as a registered nurse, the occupational health nurse should acquire a knowledge of occupational health nursing principles and practice, and keep up to date with developments in the field. As in other areas, new courses are becoming available that assist occupational health nurses to extend their skills.

Knowledge and skills in the following areas will be required:

- occupational diseases, epidemiology and toxicology
- treatment of conditions such as eye injuries, burns, skin problems, strains and sprains
- health screening procedures, including health interviews
- counselling and communication skills (oral and written)

- training and instructional skills as needed to institute specific programs
- selected laboratory procedures for biological and workplace environment monitoring (for example, to assess the need for referral to a specialist adviser)
- awareness of local community health and specialist occupational health and safety services available
- local health and safety and workers compensation legislation
- ethics (including impartiality) and privacy issues, and
- industrial hygiene.

The occupational physician or company doctor

The promotion of occupational medicine as a specialist branch is being recognised and encouraged by an expansion in the availability of appropriate training and in professional bodies (such as the Australasian Faculty of Occupational Medicine).

It is still unusual for an occupational physician to be engaged outside of large organisations. More often an organisation will employ the services of a local general practitioner on a part-time basis.

Some general practitioners are now specialising in occupational health, and the organisation should "shop around" before choosing a part-time doctor (qualifications and training of occupational physicians are discussed below). Specific knowledge of the area in which the organisation operates is a distinct advantage.

The main duties of a "company doctor" can only be outlined very generally in the following table. Many of these services would involve further referral.

Duties of a "company doctor"

1. Liaison with and supervision of the work of the occupational health service staff, including nurses.
2. Advice on policy matters.
3. Treatment and consultations.
4. Medical examinations, including pre-placement.
5. Workers compensation cases.
6. Counselling and referrals.
7. Visits and investigation of work environments.
8. Research program development.
9. Involvement in health education programs.
10. Overall monitoring of the health service's performance, including reports to management.
11. Diagnoses.
12. Prescriptive and dispensary functions.
13. Surveillance of hygiene standards, including facilities such as canteens, kitchens, child-minding centres and sanitary installations.

14. Supervision of specialised medical equipment and its use.
15. Involvement in adaptation of jobs to employees from a safety/health viewpoint, including rehabilitation programs and disabled employees.
16. Liaison with managers, and also with outside local medical services.

The occupational physician or company doctor should report directly to senior management, although it is crucial for him/her to maintain a stance of strict impartiality (that is, to avoid becoming regarded as a tool of either management or employees).

A company doctor will be of greatest benefit to both the organisation and employees if the duties of the position are interpreted to include surveillance of both the workplace (processes, materials and conditions) and workers. This, in turn, suggests it is far better for consultations to be carried out at the workplace than at an outside surgery.

Consequently, where a doctor is employed as a consultant to the organisation, it will be necessary to provide adequate medical care and consultation facilities at the workplace. Administration of these facilities would generally be the responsibility of an occupational health nurse, perhaps assisted by clerical or other staff.

Qualifications and training required

Ideally, the doctor should have some postgraduate training in the occupational medicine/health field. Among the extra areas of knowledge and skill required are the following:

- an appreciation of management styles and problems, organisational behaviour, cost/benefit analysis, industrial relations and administration
- familiarity with what employees do in their day-to-day work, including materials, plant and processes involved, and
- knowledge and experience in areas such as epidemiology, toxicology, ergonomics, safety, sanitation, health/safety legislation, health education, rehabilitation and counselling.

Under certain state legislation, provision is made for medical officers to become registered medical practitioners for specific health hazards, such as lead or asbestos.

Where particular hazards are involved, the doctor should have a thorough knowledge of the specific hazard and the appropriate treatments available, although detailed research study of the hazard may be undertaken by other service groups.

The occupational hygienist

Occupational hygiene may be defined as the detection, evaluation and control of either physical or toxic stresses in the working environment, that may have adverse effects on employee health and wellbeing. It aims to prevent adverse effects occurring, both in the short and long terms.

An occupational hygienist would be of particular value if the organisation has a health problem that needs constant monitoring, such as the use of a potentially harmful chemical or substance. Many of these are of unknown potential, so it is essential to keep levels of exposure to the lowest practical limit.

Duties of an occupational hygienist could include those in the following list.

As this list implies, the hygienist will need to possess knowledge of occupational hygiene, work study methods, sanitation, general health, research techniques and communication skills. Frequent contact with the workplace and its employees is essential, as is the need to keep abreast of new discoveries and developments, particularly the long-term effects of hazards and control measures.

The hygienist could be either a full-time employee or an outside consultant.

Duties of an occupational hygienist

1. Use established exposure standards as guides to control measures.
2. Audit the work processes/materials that involve particular hazards and develop monitoring programs to evaluate the extent of the problem.
3. Make appropriate contributions to risk management processes.
4. Work with industrial medical staff to evaluate particular tasks, processes or materials.
5. Prescribe measures for safe handling and treatment of hazardous substances.
6. Assist management to provide advice on hazards to employees, OHS personnel, compensation authorities, customers and governments.
7. Provide data for continued surveillance (that is, whether the situation is improving or worsening).

Other occupational health and safety consultants

It may be that an organisation requires the assistance of a consultant with expertise in a specific area. The occupational health and safety professionals presented in this section cover the major groups not already discussed. Because of the specialist nature of their work, all but the largest companies would be expected to engage these professionals on a consultancy basis.

Organisational psychologist

Organisational psychologists are concerned with people at work. They apply psychological principles to the study of work performance, work attitudes, working conditions and organisational structures. They aim to improve individual performance and work satisfaction, as well as promoting organisational effectiveness.

Areas of occupational health and safety with particular relevance to a psychologist include employee assistance programs, occupational rehabilitation, stress and general health maintenance.

The Australian Psychological Society is the professional body monitoring standards and representing psychologists. There is a branch and also a Board of Organisational Psychologists in each state.

Risk manager

Risk management addresses not only employee health and safety, but all elements of unnecessary costs that may affect a company's profitability, eg loss prevention, computer security.

Risk management involves minimising risk through appropriate management systems. Risk managers may look at the whole organisation, however, with respect to OHS this involves identifying hazards/risks and developing systems to minimise the risks associated with those hazards. It is a discipline oriented to practical change. (See also Chapter 3.)

The risk manager:

- identifies the hazards associated with a particular undertaking
- qualifies and quantifies the degree of risk associated with identified hazards (what potential these have to cause human and financial injury/loss)
- designs viable programs to limit exposure to the hazards/risks (so reducing the potential for injury/loss), and
- monitors and maintains these programs.

Rehabilitation counsellor

A rehabilitation counsellor provides assistance to people with problems resulting from disability or injury. In the occupational context this would refer to people who are having problems getting into, or returning to, the workforce. The counsellor may assist in the process of adjusting to the workforce through individual counselling or through coordinating various other services.

A rehabilitation counsellor will be concerned with the whole person, not simply with the person's disability. He/she may stress the remaining skills of the injured person, or may assist the worker in obtaining new skills.

Rehabilitation counsellors may also be involved in assessing the effectiveness of any return to the workforce.

Ergonomist

Ergonomics looks at people and the systems in which they work. It is about designing work (including machines and workplaces) to suit people. The employees are considered to be the chief resource at the workplace, around whom work systems should be structured.

An ergonomist considers areas such as the following:

- equipment design (handles, position of controls, etc)
- workplace layout
- environmental factors (light, noise, chemicals)
- how people handle information

- the skills people use, and
- how individual variations between people (such as height, weight, strength, intelligence) can be accommodated by a work system.

Ergonomics is a very broad science and it is hard for any one person to claim to be an "ergonomics expert". Some consultants come to ergonomics via other professions such as engineering, physiotherapy or nursing. The Human Factors and Ergonomics Society of Australia provides accreditation for its members. The NOHSC built up a large amount of information on ergonomics on its website (www.nohsc.gov.au), now also available on the ASCC's website at www.ascc.gov.au.

OHS managers — qualifications

Many large organisations employ an OHS manager (or safety, health and environment manager). When drawing up a job specification for a position of this type, either full- or part-time, the following qualities are desirable:

- knowledge of hazard identification, risk assessment and control
- thorough knowledge of relevant legislation, codes of practice and standards applicable to the workplace, including any regulations covering buildings, fire protection, electricity, vehicles, lifting appliances, certification requirements, harmful substances, etc
- a capacity to understand the nature of the organisation's activities
- other technical knowledge, such as storage and control of dangerous goods and hazardous substances, fire prevention and security
- skills in report writing and verbal communication
- the qualities of tact, patience, and the capacity for accurate, efficient handling of detail
- the ability to undertake instruction and training
- the ability to liaise with the workplace health and safety representatives and committee members
- knowledge of current developments in disease and accident prevention methods and thinking, both in Australia and overseas
- understanding accident causation and the techniques used in reporting, investigating and recording accidents and injuries, and accident statistics
- enthusiasm for health and safety work
- adaptability and willingness to undertake further training in health and safety, and to engage in outside activities such as part-time health and safety training and membership of appropriate professional organisations
- a good knowledge of the organisation's policies and procedures
- the ability to distinguish between the advisory function performed as health and safety officer and any other executive or technical functions he/she may have, and
- the ability to influence personnel, enabling change to be implemented.

The health and safety professional should be given the opportunity to acquire this knowledge and these skills and abilities. Advice on courses available should be sought from the government departments and organisations listed in Appendix A as well as university departments and TAFE colleges.

Workplace OHS services

Large companies may choose to establish an in-house occupational health service to help protect workers from occupational injury and disease. Such a service is usually staffed by occupational health nurses and/or occupational physicians. Its functions include the promotion of a healthy work environment and general health supervision, monitoring the work environment, treatment and rehabilitation, education and counselling, supervision of first aid, record keeping, research and liaison.

Location and facilities

The location, layout and facilities of an occupational health service should be planned either in conjunction with existing occupational health and safety staff or with the relevant state occupational health and safety authorities. The planning and equipping of a health centre should comply with any existing legal requirements or established codes of practice. The size of the particular business or organisation, its type of operation and, therefore, the kinds of work hazards present should also be taken into account. It should be noted that various casual, on-site services (such as audiometric testing) are available commercially.

The need for and role of such employees will be affected by the type of workplace, type of work performed, particular safety aspects, availability of community health facilities, shiftwork and other factors. If occupational health practitioners are appropriate, the following provides general guidelines on staffing recommendations:

1. Factories with over 300 employees, or shop/office with over 500 employees — one occupational health nurse and a part-time doctor.
2. Over 1,000 employees — at least two occupational health nurses and a part-time doctor.
3. Over 2,000 employees — at least one full-time doctor in addition to the nurses.
4. Over 4,000 employees — at least two doctors and at least three nurses.

Establishing a group health service

Smaller organisations may not be able to justify the expense of an on-site health service, and may wish to contract an outside service to provide some of the same functions. This may involve an initial familiarisation with the workplace, regular visits by the health service to monitor the workplace and perform health checks on workers, as well as workers and management having access to the health service at all other times. Because workers may need to attend the health service following an accident, it is desirable to engage a health service near the workplace.

Small organisations could also give thought to either maintaining a part-time service or combining with other small organisations within close proximity to establish a group health service. An obvious example is an industrial estate complex. If establishing a group health service, either a central health centre, a series of separate clinics, or a mobile centre could be considered. Several such casual, on-site services are now available on a commercial basis, offering a range of services from pre-placement testing and specific services (such as audiometric testing) to ongoing monitoring.

The suitability of any such approach will depend on the availability of nearby medical/health facilities and the needs of each organisation. It may also require frequent reviews to make sure it is continuing to meet those needs.

Problems that may have to be addressed with a group service may include greater complexity of administration (with several companies participating), a wider variety of existing and potential problems and hazards, and financial support (each company would normally pay a ''subscription'' to the service). Financial planning in advance will be very important (for example, to establish an optimum size, range of services and coverage of companies) to ensure financial viability.

INDEX

References are to paragraph (¶) number.